To the few who saw the need and the many who helped build the Museum of Flight.

A special dedication of this story is made to William E. Boeing Jr., The Boeing Company, and Bruce R. McCaw, among the most committed supporters of and generous contributors to the museum.

For Future Generations

A History of the

Museum of Flight

Howard Lovering

Documentary Media

Seattle, Washington

For Future Generations: A History of the Museum of Flight

First Edition
Printed in China

**Cover art:
Ted Huetter**

Produced by Documentary Media, LLC
books@docbooks.com
www.docbooks.com
(206) 935-9292

Team One: Bruce R. McCaw, Howard Lovering, Peggy Nuetzel, Alison Bailey,
    Craig Stewart, Aurelie McKinstry
Author: Howard Lovering
Sidebar Authors: Jay Spenser, Howard Lovering
Research/Writing/Editing: Peggy Nuetzel
Editor: Judy Gouldthorpe
Book Design: Paul Langland
Editorial Assistant: Tess Beck
Editorial Director: Petyr Beck

ISBN: 978-1-933245-42-3

Library of Congress Cataloging-in-Publication Data
Names: Lovering, Howard C., 1937- author.
Title: For future generations : a history of the Museum of Flight /
    by Howard Lovering.
Description: First edition. | Seattle : Documentary Media, [2016] |
    Includes index.
Identifiers: LCCN 2016020278 | ISBN 9781933245423 (hardcover)
Subjects: LCSH: Museum of Flight (Seattle, Wash.) | Science
    museums--Educational aspects--Washington (State)—Seattle.
Classification: LCC TL506.U6 S648 2016 | DDC 629.130074/797772—dc23
LC record available at https://lccn.loc.gov/2016020278

Contents

Foreword

Major milestones cause us to pause, reflect on the past, and celebrate the people and events that brought us to that moment in time and then to look into the future to dream about what may come next. Fifty years ago no one could have foreseen the Museum of Flight of today any more than we can envision today what will happen in the next 50 years. But now is a time to honor and thank all the men and women who created this magnificent organization.

In 2015 Mike Hallman and I were deeply honored to co-chair the museum's 50th Anniversary Gala with Honorary Chair June Boeing. Very sadly, we had lost Bill Boeing early that year. Without Bill none of this ever would have happened. A quiet force and a dear friend, he was such an amazing man, and his vision, passion, and generosity made it all a reality. We miss him tremendously. For 25 years his wife June was alongside Bill at every step moving us forward, and she continues to do so. The Gala was a huge success as we both honored Bill Boeing and celebrated our 50 years in grand style.

The final step was to chronicle this rich history. While this seemed a good idea, none of us knew what a massive undertaking it would actually be. Howard Lovering, the Museum of Flight's original Executive Director, was selected as the author, very capably assisted by his wife, Peggy Nuetzel. Howard and Peggy, principals in the firm LOGIC Inc., have worked with countless museums in the past 25 years. Their collective knowledge, contacts, and supporting files have been invaluable. The quality of their writing and editing talent will be evident in the following pages. Clearly this story was one that required both depth and accuracy. It had to convey the senses of courage, fear, panic, blind optimism, humor, and success that were interwoven in all that has transpired in this remarkable journey. I do not think we could have found anyone better to recount this saga. It has been a memorable ride and a very good one. Howard also engaged Jay Spenser to provide technical expertise, particularly in the many sidebars. Jay is no stranger to the museum, having served as MOF's curator following a previous position with the National Air and Space Museum as assistant curator. Jay has written many wonderful historical aviation books and countless articles and is a wealth of knowledge. It has been a real honor to have him contribute to this work.

So many others have likewise helped along the way, and we deeply appreciate all of them. Other vital members of this team included MOF's superhero for over three decades, Alison Bailey, and our publisher, Petyr Beck of Documentary Media. Petyr and his associates have provided wisdom, guidance, enthusiasm, support, and patience. Amy Heidrick and the Museum's library and archives group, as well as Mike Lombardi at Boeing, helped us find the photographs and documents needed to make this book come alive. Craig Stewart and Aurelie McKinstry of Apex Foundation have been invaluable in the organizing and financing of the project.

Recounting the story of the Museum of Flight, from the early days of PNAHF (Pacific Northwest Aviation Historical Foundation) to the present, has been quite the challenge. Even now it is difficult to comprehend all that has been achieved to create this wonderful institution. Countless individuals and businesses, volunteers, supporters, donors, employees, management, and the board have played crucial roles in this story. Though we were not able to give them all the recognition they deserve in this book, their contributions are deeply appreciated. As memories often differ, records conflict or do not exist, and many key individuals are no longer with us, our goal has been to be as complete and accurate as possible. While we undoubtedly missed things, everyone who has been part of this fabulous museum should be immensely proud!

Perhaps it has been fate, but somehow I have been both a witness to as well as a participant in much of this story. In high school I knew or knew of so many aviation legends, particularly those from Boeing. Then they were my classmates' parents, neighbors, or just local heroes in a much smaller Seattle community where everyone seemed to know each other. As a fledgling pilot in 1967, I found myself being immersed in the world of aviation. When the new Seattle Center exhibit opened, it became a favorite of mine. Less than a year later I was working at Paine Field just across the airport from the massive new Boeing 747 plant. RA-001, the world's first jumbo jet, was being constructed there. I was present for the rollout and saw it take to the sky for the first time. Never could I have imagined that I would connect with this historic aircraft decades later.

At Paine I met two unforgettable characters, MOF co-founder Jack Leffler and author Ernie Gann, flying the 247. Both would become lifelong friends. I got to know others, including the likes of Clayton Scott (both of them), Brien Wygle, Chuck Lyford, Paul Bennett, Lew Wallick, Don Filer, Bob Hoover, and members of the Blue Angels team, to name a few—people whose paths I would cross again and again. They helped me understand the importance of preserving the rich history of the earlier days of aviation. Bill Boeing Jr. and I became acquainted, along with his daughters Bebe and Gretchen. A few years later Gretchen introduced me to Howard Lovering and persuaded me to get involved and help out in the early days of MOF. Some 38 years later I still am. Thank you, Gretchen!

The Museum of Flight inspires, educates, and entertains both young and old. It is about the human spirit and it is ageless. Our education programs are second to none. The site itself is historic, situated on Boeing Field, an active airfield where the first flight in Seattle took place in 1910. The museum hosted the aviation celebration for the Washington State Centennial, the 100th-birthday party of the global Boeing Company, and many historic flights including the Concorde "Flights to Nowhere," "Friendship One," and many, many others.

Starting with the iconic "Red Barn" building itself, the collection touches on and spotlights so many moments in history. The museum's Library and Archives include the letter from W. E. Boeing Sr. hiring his first engineer, Wong Tsu, who had just graduated from MIT in 1916, the Wright Company Papers, and Captain Elrey Jeppesen's personal "little black book." Aircraft include the Boeing & Westervelt replica, the first and only Caproni Ca.20 fighter, the early Boeing Models 40B, 80A, 247, B-17, and B-47, together with the fastest Blackbird, the beautiful Concorde, as well as RA-001, "Queen of the Skies," the massive first-ever 747 that would change the world.

The robust collection also holds Deke Slayton's tiny astronaut pin, moon rocks, Pete Conrad's space suit checklist from the Moon, and pieces of the Apollo F-1 engines that launched that same flight and only recently were recovered from the depths of the Atlantic Ocean by Bezos Expeditions. The modern space collection highlighted by the Space Shuttle Trainer in the Simonyi Space Gallery continues to grow. Both the campus and the mission have expanded far beyond the wildest dreams and plans of the founders. However well our visitors know this place, there are new surprises to be found on each and every visit.

Fifty years is suddenly 51, and we are already well on our way to 100. We hope you enjoy these memories and that you will all be part of the chapters yet to be written.

— Bruce R. McCaw

Now fully restored, this was the only airworthy DC-2 in the world when Joe Clark, Clay Lacy, and Bruce McCaw acquired it for the Museum of Flight.

Preface

The Museum of Flight is a great museum—entertaining, popular, a valuable community resource. It is a remarkable transportation museum and educational platform that inspires other institutions.

Typical models for creating great museums most often emerge from the interests of the wealthy or vast resources of governments. It is tantalizing to read about the Machiavellian machinations sometimes involved in acquiring an addition to a great art institution; the detective work of scholarly curators; global travel and adventure; application of huge sums of money from wealthy donors; stealthy negotiations; the intervention of senators, even presidents and prime ministers; and finally the glorious moment when a precious collection and new wing are premiered with the general public. Reading a museum book such as *Making the Mummies Dance*, by Thomas Hoving, about the expansion of the Metropolitan Museum of Art in New York, is to some of us as fascinating as a James Bond movie. The intrigue, the money involved, the patrician characters, not to mention the romance and fine dining, are extravagant.

The government model is compelling, with access to capital and well-situated properties in its land holdings. In addition, with various public laws, governmental institutions can reach out without competition for the artifacts of history in their domain. This applies in particular to aerospace, with its many treasures in the ownership of the military or NASA or some other federal agency, building an attraction that is predetermined to be sustainable. Neither of these models pertains to the Museum of Flight.

A common perception is that the institution is the "Boeing" Museum of Flight, sprung full-blown from the actions of a small group of community leaders with the leverage and big check written by The Boeing Company. This is simply not accurate. The real story is more complicated, vital, entertaining, instructional, and ultimately inspiring.

Building the Museum of Flight was not the work of a few founders. It is better characterized as the triumph of a group effort. What started small grew to be a large team, recruiting thousands of supporters, building from the bottom up an institution that not only shares with but also belongs to the visitor. The cast of builders was diverse. Enthusiasts brought a host of ideas and a lot of attitudes. Together, these disparate views and interests have powered the new institution.

The Museum of Flight hums along from early morning until often late at night. It daily accommodates more than a thousand visitors, arriving from all parts of the country and abroad. The museum educates, entertains, and inspires young and old alike with a magnificent collection of aerospace artifacts, with the rich interpretation of themes, and with what are considered the most extensive flight history and technology programs in the world. It is much respected, considered an institution that has just been there forever—and for its younger fans, it has been.

Some of the most iconic structures in the world, once built, reshape their human environment. They insinuate themselves into the geography in such a way that they come to define the community. All great cities have their icons, and Seattle has a few; the Museum of Flight has joined that company.

The narrative of the first 50 years—the dreaming, hard work, setbacks, accomplishments, the unwillingness to quit—is much more compelling than a perception that the museum has always been there or that building it was easy. The Museum of Flight had as great a chance of failure as of success.

The story of the museum cannot be told definitively at any point, as it is moving and growing so rapidly. What now occurs each month at the museum in attendance, programs for all ages, special events, additions to the archives, collection, and facilities surpasses what happened in a year or more during the formative years. The pace of operations is so fast, the scale so large, that only a daily diary can record the progress. The best we can do in this encapsulation is to discuss major themes and the thread of growth, the contributions of leadership, and the profile that makes this institution distinct while emphasizing the human drama. This is just the beginning of an outline for a continuing conversation with history. As one who was there, I can assure you it is a privilege to participate in telling this story.

— Howard Lovering

Left: This Boeing
80A had seen better
days when rescued
from an Alaskan
garbage dump.

Right: An early
PNAHF logo.

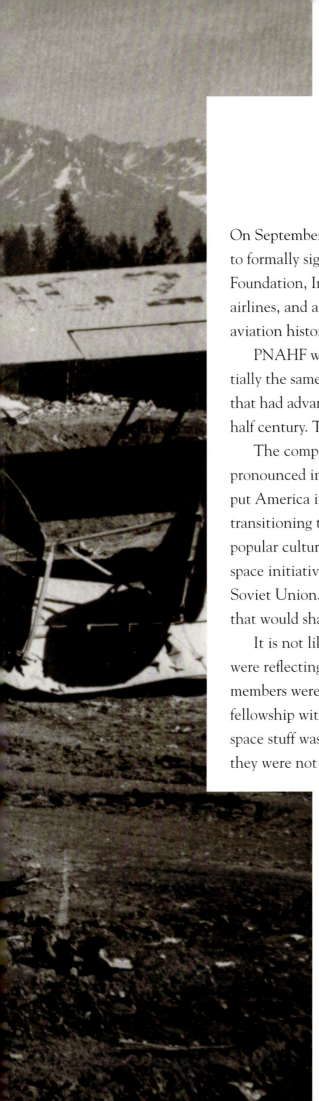

Introduction

On September 14, 1965, twelve gentlemen met at the Greater Renton Chamber of Commerce to formally sign the incorporation papers for the Pacific Northwest Aviation Historical Foundation, Inc. (PNAHF). These founding members worked at The Boeing Company, airlines, and a variety of businesses, but were joined by a common interest in preserving aviation history.

PNAHF was like so many other organizations springing up around the country for essentially the same reason. Aviation, now encompassing aerospace as well, was a fast-paced industry that had advanced to a point of changing global travel, economies, and cultures in just over a half century. The history, so young, was largely unrecorded, and its major artifacts were at risk.

The comparatively new aerospace industry was in a major period of growth, spurred by the pronounced initiative to send an American to the Moon. The result was an aggressive effort to put America into a leadership role in space exploration, and many aviation companies were transitioning to aerospace. Astronauts were the epitome of American heroes, and much of the popular culture became decorated with space themes. This was a busy time for the NASA space initiative, but it was more than just a governmental program of competition with the Soviet Union. It was the beginning of a new industry, with an acceleration of technologies that would shape the rest of the century and beyond.

It is not likely that the gentlemen at the Renton Chamber were reflecting much on this space program, although several members were commercial test pilots who had a personal fellowship with the Mercury 7, and all had an interest. This space stuff was new, and the machines were not artifacts and they were not yet in danger of being lost.

Top: The Boeing 80A in its heyday at Burbank airport, circa 1930s.

Bottom: Executive Director Howard Lovering with Peggy Nuetzel at the opening gala for the Great Gallery, 1987.

For the founders of PNAHF, there was an urgent motivation to save aviation artifacts and to care for a big new acquisition, the venerable Boeing 80A biplane transport, a valuable basket case that needed attention. But no matter how instrumental it had been in forging a new global industry, this old Boeing commercial biplane could not muster much attention given the new rockets and dreams of walking on the Moon. Another acquisition was hinted at—a bit of a mystery at the time and more reason to incorporate their interests and intents—but there was the 80A to attend to, other stuff to save, a property for a museum to be found, and time was wasting.

None of the PNAHF founders could have had a glimmer of understanding about how far their idea could go.

Twenty-two years after the founders formally convened, the new Museum of Flight celebrated its opening at Boeing Field. On a temperate morning, hundreds of students disembarked from a fleet of yellow buses to join the growing crowd gathering around the temporary stage outside. The ceremony and ribbon cutting were modest if that is possible in the company of Mercury astronauts, the governor of Washington, CEO of the world's largest aerospace company, and the Vice President of the United States. That evening, as guests arrived in tuxedos and gowns, they knew the moment was exceptional and historic as they strolled to the celebration for the grand opening of the Great Gallery of the Museum of Flight.

As if in respect for all the hard work, the frustrations, the setbacks, the good reasons to quit, even the formal opposition, this new museum was happening, was opening, and it was stunning.

For those who had endured these two decades of hard work and frustration, it was a beautiful occasion. To quote Winston Churchill, "This is not the end. It is not even the beginning of the end. But it is, perhaps, the end of the beginning."

On November 19, 2015, the Museum of Flight announced that it would be the recipient of remains of an F-1 booster engine from Apollo 12, a mission that was the second to land American

Jeff Bezos presents an F-1 booster engine from Apollo 12 to the museum.

astronauts on the Moon in 1969. This artifact was retrieved from a depth of 14,000 feet in the Atlantic, where it had been jettisoned 46 years earlier. The founder of Amazon, a company headquartered in Seattle, had sponsored the retrieval of these Apollo parts from the depths.

It was just another day at the Museum of Flight, but it was a milestone that put an exclamation mark on the day 50 years earlier and the dream of building a museum. The museum had been imagined at the same time that the Apollo program was proving American technological leadership. The museum idea grew slowly, as the Apollo 12 booster engines rested on the Atlantic floor. Grandest notions did not envision a museum opening with Mercury 7 astronauts in attendance, nor that 50 years later this museum would be the first public institution to display parts of the massive Apollo booster engine that had pushed astronauts to the Moon.

Over five decades, the Museum of Flight has grown from a few dreamers with a battered commercial biplane aircraft to a world leader in aerospace education. No one now can easily imagine what might come along over the next 50 years. But all can be certain that the Museum of Flight will continue to push the envelope for future generations.

The Museum of Flight campus reflects the evolution of the age of flight.

Left: A Boeing 80A flies above Seattle's Ship Canal and Lake Union during the 1930s.

Right: The Boeing B&W was the company's first product, seen here on Lake Union, Seattle, in 1916.

CHAPTER ONE

Some Day We Have to Get Organized

Seattle's Museum of Flight and the commitment to save and interpret Pacific Northwest aviation history began with a discovery in an Alaskan dump. That archaeological find, the vision of a few fascinating characters, and years of dogged work resulted in one of the finest flight museums on Earth.

Of course, Seattle is a recognized aviation community. Its affair with flight began shortly after the turn of the 20th century. Though far from the center of the action, William E. Boeing and associates built a company that would become a global aerospace leader. From early demonstration flights to jumbo jets and into space, Seattle and The Boeing Company have emerged as key players in the region and the world.

The first Boeing aircraft, the B&W seaplane, took flight in June 1916, not much more than a dozen years after the Wright brothers' first powered flight on the other side of the country. Fifty years later, with so much aviation progress, there was little more than a paper trail documenting the early history of the Pacific Northwest.

During the 1960s, aviation buffs around the country had begun to express concern over the loss of vintage aircraft and articles of flight history. Some regional groups organized to save these items and some talked of building air museums. In Washington, D.C., the Smithsonian Institution had just embarked on serious plans for a new National Air and Space Museum in a place of national significance. The flight museum sector was emerging.

In Seattle, a group of enthusiasts was already gathering around these issues. It was the treasure in the dump that booted up the program.

On May 26, 1960, the *Anchorage Daily Times* carried a story of the sad ending for a historic Boeing 80A aircraft. After its years of use in hauling equipment and materials for construction in remote locations, the workhorse was wrung out and finally dispatched to its inglorious retirement at the Anchorage landfill.

The Boeing 80A was an early passenger airliner built for safe and comfortable travel and operated by Boeing Air Transport, a forerunner of United Airlines. Nicknamed the "Pioneer Pullman of the Air" because of its railcar-sized accommodations, it carried 12 to 18 passengers in different configurations. Flying between the West Coast and Chicago from 1929 until 1934, the aircraft cruised at 120 miles an hour, with amenities that included hot and cold water. The 80A introduced the first women flight attendants, a group of trained nurses called "stewardesses."

Replaced in the Boeing Transport commercial fleet with new, larger aircraft, the 80A was sold in 1941 to Morrison-Knudsen Construction and stripped to haul cargo in Alaska during World War II. After the war, it was again set

Top: The venerable 80A, nicknamed "Yellow Peril," in an Alaskan landfill in 1960.

Bottom: In 1930, Boeing Air Transport introduced the first "stewardesses," seen here with the 80A.

aside and then given to Robert C. Reeve, a famous Alaskan bush pilot, who had flown this particular 80A for the construction company during the war.

Reeve, the founder of Reeve Aleutian Airways, had an operator's reverence for this aircraft and its rugged performance over years of hauling tons of freight to outlying stations. "The biggest single piece Reeve ever flew in the plane was an 18,000-pound boiler," reported the news article. "This was so big he cut out the side of his plane in order to put the boiler in and then put the side of the plane back on again."

According to Tilly Reeve in Peter M. Bowers' book *Flying the Boeing Model 80*, she first saw the plane, registration number NC-224M, in 1941 in Nabesna, Alaska, where her husband Bob was transporting workmen and materials into Northway for the construction of an airport. "One day our four-year-old son came running into the tent saying, 'The fighter pilots are coming to see us again—I can hear three motors.' We ran out to watch them land when instead we saw a yellow

behemoth descending over the trees. What a monster—
with three engines, three tails, and four wings. It was Bob
bringing in the 80A."

Reeve commented affectionately on the aircraft in an
April 1954 edition of *Aero Digest* reporting the retirement of this 80A. "She's in her original condition except
for the door and the girder, and of course the inside is
freight car style. We don't know how many times she has
been recovered, but we find her a very fine airplane, very
modest in maintenance, has a short landing and good
takeoff and we like it."

After its service, the 80A sat derelict at Merrill Field
in Anchorage from 1947 to 1960, where, according to Tilly Reeve, the
"224M was a playhouse for our two younger sons." In a letter to the
museum in 1985, Bob and Tilly's son David Reeve remembered, "My
little brother, Whitham, and I spent many an hour 'flying' the 80A all
over the European skies. We bombed many a Hun and shot down more
Stukas and Messerschmitts than were produced by Germany during the
entire war. I must admit we were shot down a couple of times ourselves.
Being the oldest and therefore having proprietary rights to the 'left seat,'
I was always the Captain and my little brother the Co-pilot. He was
always outright killed in any battle, while, mortally wounded, I managed
to crash land our flaming aircraft. Many a time I bailed out of the forward
right-hand door, only to be slashed to bits by the right prop or impaled on
the horizontal stabilizer. Of course, this was true only as long as we did
not enter that 'dreaded flat spin.' "

Reeve had offered the historic aircraft to the Smithsonian for their
new air museum, but that deal fell through when the institution realized it could
not afford the cost of the relic's transportation from Anchorage to the newly
proposed museum on the Mall in Washington, D.C. The 80A was relegated to
the sidelines, where it sat outside Reeve's hangar until it was moved to the Anchorage
landfill in 1960.

Harriss Darby, chief photographer for the *Anchorage Daily Times* and an aviation enthusiast,
launched an effort to rescue the orphaned aircraft from the landfill. Realizing its value, he
purchased it for 50 dollars from the City of Anchorage and moved it out of harm's way.

Top: A boiler is
loaded into the 80A
for transport via a
large hole cut into
the fuselage.

Bottom: In this 1954
Aero Digest, the 80A
was characterized as
an "ageless veteran."

Boeing 80A

The world's only remaining Boeing 80A resides at the Museum of Flight. Accorded a place of honor in the Great Gallery, this 1929 biplane is a crown jewel of the museum's collection because it is the earliest surviving Boeing airliner designed for passenger-only travel.

The United States was a far different place in the "roaring twenties," a frenetic era of jazz, flappers, art deco design, burgeoning industry, and growing scientific prowess. Back then, people took the train and it required more than three days to travel from coast to coast.

Europe, with its closely spaced capitals, enjoyed passenger air services throughout the 1920s. In contrast, North America's vast scale and challenging geography meant that U.S. commercial flight operations began with air mail, not passengers. Thanks to technological progress, however, the stage was set by the latter 1920s for passenger operations to emerge here as well.

Fokker and Ford trimotor airliners were already in U.S. service when Boeing developed the Model 80, its first true passenger airliner. The company was then building Model 40 mail planes, the latest version of which could also carry up to four paying passengers. But air mail had priority, and passengers were apt to be left behind if too many mailbags awaited loading.

In contrast, the new Boeing Model 80 was designed specifically for passengers. Its inviting cabin was heated and featured wood veneer, leather seats, shaded reading lamps, hat clips, and controllable air vents at each window. A rear lavatory offered hot and cold running water. Despite these amenities, however, the Model 80—like all early airliners—was noisy, bumpy, and uncomfortable because planes back then were unpressurized and flew through the weather rather than above it.

Like the Ford and Fokker airliners, the Boeing Model 80 was a trimotor. Unlike those types, it was a biplane rather than a monoplane. The extra wing area was needed for greater lift to safely clear the Rocky Mountains on Boeing Air Transport's Chicago–West Coast route. The 12-passenger Model 80 entered service in September 1928, followed one year later by the improved 18-passenger Model 80A, with more powerful engines.

In 1931, Boeing Air Transport became United Air Lines, ancestor of today's United Airlines. The previous year, the carrier and its trimotor Boeings had made history by introducing the world's first stewardesses, all of whom were then registered nurses. It was the idea of Ellen Church, a young nurse and private pilot, who felt that the presence of female flight attendants would help aviation gain acceptance by reassuring nervous fliers. On May 15, 1930, Church herself put this idea into practice on a 20-hour flight between San Francisco and Chicago, with 13 stops in all.

Thousands of Boeing jetliners today whisk people and cargoes swiftly to all corners of the globe. These tireless workhorses—aerial ambassadors linking people and cultures—promote understanding, commerce, and the free flow of human ideas and energy. So indispensable are they to our 21st-century infrastructure that we often take them for granted. As the Museum of Flight reminds us, it all started humbly with magnificent early transports like the Boeing 80A.

The city of Seattle, with a Boeing 80A flying above and the Smith Tower skyscraper anchoring the skyline. Both were symbols of a new age.

The 80A was dismantled in Alaska (top) and then packed into a shed.

As recounted in the book *Flying the Boeing Model 80*, Darby "had to go as high as the mayor to get a hold order on the normal dump procedure of burying junk within a week of receival. He needed time to organize the salvage operation. He and some friends disassembled the plane on the dump and transported it to a shed five miles from the airport, where it was put under cover with the remains of some other antique aircraft that he had acquired."

Darby immediately wrote to Boeing's public relations department in Seattle, hoping to solicit the company's interest in the plane and asking for any photos or drawings that could aid in its restoration. In his letter of May 29, 1960, he detailed his adventures in the landfill saying, "he felt like the flea that married the elephant. Acres and acres of it, and it's all mine!"

At Boeing, Harl V. Brackin Jr. was assigned the task of locating blueprints and other documents relating to the aircraft. Brackin had read about this particular 80A and its Alaskan history in the *Aero Digest* article, in which Bob Reeve had avowed, "Here's an airplane that certainly should someday be in a museum."

As archivist and historian at The Boeing Company, Harl practiced his abiding love of flight history in the corporate setting, while nurturing the dream of a public museum. A humble, soft-spoken man, he is remembered as always smiling as he graciously accommodated requests for archival information. Born with limited hearing, Harl had a pronounced lisp. He was a thoughtful and friendly gentleman, dogged in his efforts to preserve aviation history.

In July 1961, Brackin received another letter from Darby in which he lamented his failure to take on the massive restoration necessary. He regretfully informed Brackin that the aircraft was up for grabs. Brackin did some sleuthing and found that a California museum was interested in acquiring the artifact and realized it was time for action if Seattle wanted to get into the game.

In parallel, United Airlines Captain Jack Leffler had also heard about the 80A from his friend Pete Bowers, a Seattle aviation historian. Leffler was Brackin's polar opposite. Harl was tall, Jack short, Harl of courtly manners, Jack a practical joker. Harl was into books and Captain Jack was an adventurer. Despite all of these perceived distinctions, they were a complementary couple. With Captain Jack playing the colorful role of a pirate, Harl could be identified under the flag of the privateer, the administrator assigned his duty, but also willing to go outside the rules. Together, they plundered with the best.

Captain Jack was a military-trained glider and airplane pilot who flew for United Airlines from the era of propeller aircraft well into the jumbo jet age.

It's not clear if Captain Jack was much of a fan of flight museums—probably he was not. Most museum collections had static planes on display, and he believed that vintage aircraft should fly. But he was unconditionally dedicated to preserving these significant artifacts, particularly those that had such influence in the region. It was no surprise that he tracked down the Boeing 80A.

He was portly, not likely to be picked from the crowd as someone

Top: Jack Leffler (right) standing with Harl Brackin, holding a model of the 80A.

with the right stuff. Profiled in a June 1966 *United Mainliner* magazine, Jack was referred to as "a legend of the flying fraternity, one who added gusto and color to the profession." His intelligent humor and compelling personality made it easy for him to forge relationships at all levels. His friendship with and affection for Eddie Carlson, the legendary hotel man who became president and chief executive of United Airlines' parent company, UAL Incorporated, proved his comfort with leadership, his gravitas, and the human equity he was amassing. Jack called his powerful friend "Fast Eddie." Eddie Carlson, as far as anyone can remember, called Jack "Captain," usually framed with a broad smile.

Those who knew Leffler remembered his humorous storytelling, and his interest in art, photography, and music, as well as his passion for aircraft and aviation history. He was an accomplished cartoonist, a model-airplane builder, and a vintage-auto collector.

Captain Jack contacted Darby and Redden, and during one of his flights north in August 1963 acquired the aircraft, the terms not a matter of record. The plane was in terrible condition, with only the main fuselage section intact. Something remained of the wings, and a few assorted components had been dug from the snow, but the treasure was now in the hands of Captain Jack. It was akin to finding a pearl while dumpster diving.

Once the aircraft was identified as authentic, Captain Jack introduced the idea of restoring the 80A. His suggestion was to move the plane to Seattle as a tangible asset to prove the value of a flight museum.

Harl V. Brackin Jr.

Harl V. Brackin Jr. circa 1970s.

In 1942, a young engineer fresh out of college went to work for Boeing. Harl Brackin was assigned to the company's wind tunnel, where he contributed to programs like the B-17 and B-29. As World War II wound down, he helped evaluate captured German data showing the value of wing sweep to high-speed flight. This work contributed to the revolutionary B-47, the first large swept-wing jet ever produced.

Harl Brackin's job fascinated him, but then so did all human activity in the field of flight. He realized it was crucial to document and preserve Northwest aviation history before it was lost. His zeal and spare-time efforts in this regard eventually evolved into an informal but substantial company archive. Boeing made it official in 1962, appointing Harl its first-ever archivist/historian.

Often in collaboration with fellow Boeing employee Peter Bowers, Brackin followed up lead after lead to secure documents, photographs, films, and artifacts that might otherwise have been discarded. Using his growing archives, he was masterful at tracking down drawings, technical data, and other information needed by Boeing. The rest of his time was spent answering queries from the general public.

Showing his kind and supportive nature, when a boy whose father had flown Flying Fortresses during the war wrote requesting an actual B-17 pilot's manual, Harl instead sent him a photocopy of one, with a cover letter that began, "I'm sorry I can't take as good care of you as the B-17 took of your father." As a public face of the company, he was among its best ambassadors.

In September 1963, a letter arrived out of the blue from United Airlines Captain Jack Leffler. The airline history buff, a Seattle-area resident, wrote to say he had tracked down and purchased what remained of the last surviving Boeing 80A trimotor. The vintage 1930s airliner was in Anchorage, Alaska, and Leffler was offering to donate it if it could be transported and housed for eventual restoration.

It was high time the Pacific Northwest had an aviation museum, and here was a golden opportunity. With this in mind, Harl, with help from retired Boeing Chairman Claire Egtvedt, drafted the charter for the Pacific Northwest Aviation Historical Foundation (PNAHF) in 1964. Starting with no money but a fortune in dreams and enthusiasm, this volunteer nonprofit organization would evolve over time into today's Museum of Flight Foundation.

The basket-case 80A was duly acquired, and volunteers set to work restoring it. Meantime, Harl persuaded the City of Seattle to provide free space at Seattle Center for an embryonic museum. "We can do it!" he always said when describing his vision for what the real museum might someday be.

Harl is best known for saving the Red Barn, Boeing's original factory. In 1975, he convinced Boeing Chairman T. Wilson not just to donate the threatened structure to the PNAHF, but also to help with moving it and restoring it. With so much else on his plate, Wilson initially thought the idea little short of crazy, but in 1997 he publicly thanked Harl for his vision and tenacity in making it all happen.

Resplendently restored, the Red Barn opened in 1983 as the Museum of Flight's first building on Boeing Field. Sadly, Harl Brackin did not live to see it. The energetic dreamer died unexpectedly in his late fifties in 1977.

Brackin working on several models of what would become the Museum of Flight.

When Captain Jack delivered the news to an elated Brackin, the 80A restoration ball began to roll. Brackin suggested a presentation to the Boeing Management Association (BMA), an employee group that was influential enough to help gain their company's support of the project. In November 1963, Brackin and Leffler challenged the BMA to assemble a team from their membership to restore the Boeing aircraft. This deal cemented a valuable partnership.

On a chance encounter in Seattle, Captain Jack pitched the 80A recovery and restoration project to the legendary Boeing chair William M. Allen and was delighted at his enthusiastic response. The Captain also enlisted support from the Washington State Historical Society and the senior United States Air Force officers at nearby McChord Air Force Base. Captain Jack and Harl conspired to find a way to transport the 80A from Alaska to Seattle with BMA and Air Force support.

Unloading the 80A from the belly of a C-124 Globemaster II in February 1964.

In February 1964, the U.S. Department of Defense approved a Military Air Transport Service airlift of the 80A from Elmendorf Air Force Base in Anchorage to McChord Air Force Base, near Tacoma. Late in the afternoon of February 25, 1964, a Douglas C-124 Globemaster took off from Elmendorf, carrying the fuselage and small parts of the 80A, later cataloged as the Foundation's first artifact, NC-224M. What was left of the wings and other parts were to arrive on another C-124 flight in March.

Local media, Captain Jack, and a group of Boeing employees met the flight at McChord. The aircraft was trucked to temporary storage at Boeing Plant 2, across the street from Boeing Field. The deal was done, the goods delivered, but the immensity of the obligation had not yet settled in.

The restoration cost would be steep whether the plane was flyable or used for static display. Pieces were missing, its wing struts were broken or bent, and the fabric and corroded metal parts required replacement. This was the kind of gift that would keep on taking. An aircraft restoration survey team estimated a materials-only budget at $30,000 to $45,000, optimistic in amount but a far reach for a small group with no money.

Nevertheless, the aircraft represented a priceless asset that was relevant to the industry and the

Leffler (center), Harl Brackin (far right), and BMA members greet the 80A on arrival at Boeing Plant 2.

community. It was the only remaining 80A from the small production run of 10. Captain Jack's acquisition was one of opportunity and challenge. In one photograph Captain Jack and representatives of BMA are looking at the desolate fuselage strapped to a lowboy as it arrived at Boeing Field, their faces reflecting their most-likely thoughts: "What have we gotten ourselves into and where do we go from here?"

According to Brackin in a *Seattle Times* interview regarding the 80A acquisition, "The number 80 played a magic tune with this plane. It was called the 80; it cost $80,000; and the wingspan is 80 feet." The reporter added another 80 with a few more zeros, writing, "The cost for restoring the plane to airworthiness is $800,000."

In August 1964, struggling with the task of caring for the 80A, Brackin created and presented an audiovisual pitch to a group that included Captain Jack; recently retired chief Boeing test pilot Albert Elliott Merrill; Renton Aviation owner B. Ray Pepka; director of the Washington State Aeronautics Commission at Boeing Field William Gebenini; Northwestern Mutual Insurance agent and amateur historian Robert Hitchman; Boeing executive Walter Kee; and chairman of The Boeing Company Clairmont L. Egtvedt. They agreed to organize the Pacific Northwest Aviation Historical Foundation and establish its mission to preserve regional flight history. The time was right, and with encouragement and legal assistance from Egtvedt, Brackin hurried the necessary paperwork.

On September 14, 1965, the founders met at the Greater Renton Chamber of Commerce office and formally signed the incorporation documents for Pacific Northwest Aviation Historical Foundation, Inc. (PNAHF). Attendees included PNAHF's first elected president, test pilot Elliott Merrill.

Clairmont L. Egtvedt

Claire Egtvedt worked for Boeing from 1917 to 1966 and helped draft the charter of PNAHF.

Claire Egtvedt became Boeing president in 1933 and chairman of the board the following year. Had he not taken up the airplane company's business reins, its war-winning Boeing B-17 Flying Fortress would not have existed to play a crucial role in World War II.

Born in 1892 on a farm near Stoughton, Wisconsin, Claire moved with his family to Seattle in 1911. The forthright and energetic youth attended the University of Washington, where he majored in mechanical engineering. After graduation, he joined Boeing in June 1917 as an engineer and draftsman.

The United States had just entered World War I. Boeing, then just one year old, was a hotbed of activity in support of the nation's war effort. After the armistice of November 1918, however, military contracts were summarily canceled and the company's employment fell from 282 people to just 67. For Egtvedt, it was a sobering early lesson in the highly cyclical nature of the airplane business.

"I think I can build a better airplane," William E. Boeing had said before launching his company. Boeing was one of more than 300 aviation firms formed before or during WWI across the United States. Of these, just one in 10 ever achieved production, and none aside from Boeing survives to this day. Most fell victim to the Great Depression, which bottomed out in the early 1930s. But Boeing did well thanks to its founder's commitment to excellence. "Let no improvement in flying and flying equipment pass us by," he'd said in 1929. Claire Egtvedt shared this conviction.

Bill Boeing announced in 1934 that he was leaving the aviation business, dismaying the company that bore his name. Fortunately, there was at least no question as to who should succeed him. Claire Egtvedt, the fast-rising star, was the obvious choice.

Just two years after arriving at Boeing, Egtvedt had been named chief engineer. By 1926, he was a vice president and its general manager. Throughout this rapid ascent, he threw himself heart and soul into learning every aspect of the business. Among his favorite duties was that of flight-test engineer, when he'd strap in to monitor and record all aspects of airplane performance during test flights.

Egtvedt helped create the Boeing 247, which flew in February 1933. That August, he was promoted to president of Boeing, which then also manufactured single-seat military fighter planes. His first major act was to redirect the company to specialize in large aircraft. Thanks to this new focus, Boeing in the 1930s created the Model 307 Stratoliner, the world's first pressurized airliner, and the Model 314 Clipper, biggest and best of Pan American Airways' famous ocean-conquering flying boats.

But before those types flew, the Model 299 took wing in July 1935. This gleaming prototype— then the largest land plane in the world—embodied Egtvedt's personal vision of a flying dreadnought or aerial battleship suitable for long-range bombardment. With typical courage, he convinced Boeing to gamble enormous internal resources to develop what evolved into the B-17 just in time for WWII.

Egtvedt served as Boeing chairman for many years after relinquishing the day-to-day responsibilities of running the company. During this period, he saw his decision to refocus on large-airplane programs vindicated by the success of the B-29, B-52, C-97, 707 and its military cousins, and 727.

Passionate about aviation history, Egtvedt shared the vision of a flight museum for Seattle. He met with Harl Brackin and others in 1964 to define and draft the charter documents that created the Pacific Northwest Aviation Historical Foundation. Thereafter, he remained available to provide further expertise and support as needed.

Claire Egtvedt retired from Boeing in April 1966. Before he died at age 83 in 1975, he saw the biggest Boeing airplane of them all, the 747 jumbo jet, become a global success.

Top: The Museum of Flight board and staff visit the Boeing 80A restoration project at Boeing's Auburn facility.

Bottom: Signature page from the Pacific Northwest Aviation Historical Foundation Articles of Incorporation.

Others on the founding board included Brackin, Pepka, and Hitchman, who had been at the initial presentation, and Seattle First National banker Donald Lindsey, West Coast Airlines vice president Tom Croson, Richard Matheson, who was a special assistant to the chairman of the board of United Airlines, and Harold "Kit" Carson, another test pilot whose career included a stint as part of Boeing's commercial airplane division. C. Don Filer, who had hosted and joined some initial meetings, later bemoaned the fact that he missed the Foundation's inaugural meeting because he wrote down the wrong address.

The members represented aerospace and local business management, and they all seemed eager to promote the urgency of saving historic aircraft to anyone they felt could help the Foundation's cause. Their exuberance, their boldness, became a characteristic of the organization.

PNAHF was incorporated as a category 501(c)(3) nonprofit, tax-exempt entity, and chartered for educational, historical, and scientific purposes. Its Articles of Incorporation and by-laws were filed with the Washington Secretary of State on September 20, 1965, and the initial U.S. Treasury exempt-from-taxation letter was received on November 30, 1965. The new aviation foundation was formally established and prepared for action.

The nonprofit designation made it possible to solicit funds and in-kind contributions, as well as receive loans of artifacts. Members of PNAHF began at once to write letters to the military, governmental agencies, other museums, and potential partners to develop a collection for the museum. With the nonprofit startup, there was a license to deal, and the group wasted no time.

The Foundation's initial headquarters was at 300 Rainier Avenue North, Renton, a space made available by the Renton Chamber of Commerce. The Foundation's name was long, and people had to grapple with how to use it in a short sentence. Almost no one could remember what the acronym designated, and PNAHF simply became P-Noff, or for the uninitiated, P-Naff. Even the founders were not sure of the pronunciation. It would be a decade before leaders adopted the simpler Museum of Flight Foundation title, but that is another story.

With his deep appreciation for Boeing aircraft, Elliott Merrill was a popular choice as president. Brackin continued to push for document retention and discovery, and Captain Jack had just what he needed, a structure that would allow him to track artifacts and cut deals. The board fell in behind this leadership.

Brackin was a devout corporate historian, and a man consumed by the fear of losing valuable artifacts. His volunteer role was not part of his job description; in fact, it could have been a distraction. But he managed to walk that line, never missing an opportunity to pitch the idea of an air museum to any executive who happened to pass through.

Early membership card featuring the Boeing 80A.

The early encouragement from Claire Egtvedt and his successor Bill Allen, as well as the up-front support from BMA, sent out a signal that the aerospace industry was willing to get behind this project. Brackin envisioned an important institution that attracted and served the general public, and he wanted to become its curator. He enlisted the help of hobbyists and collectors of aviation photographs and memorabilia, and people who worked in the industry, including Boeing engineers, aircraft designers, and test pilots.

The Foundation began to shape its image as a dedicated and growing influence. News articles cropped up in support of its activities; early musings appeared on the prospects of building a museum, and more aircraft were identified as targets for acquisition.

Boeing Company volunteers designed the PNAHF logo and associated artwork for the letterhead and printed materials. PNAHF's founding brochure included the vast geographic area of Washington, Oregon, Idaho, Montana, Wyoming, and Alaska, with the largest sector encompassing the Northwest Canadian provinces of Saskatchewan, Alberta, British Columbia, and the Northwest and Yukon Territories.

Including Alaska made sense—the 80A was found in that state, which had some ties with Seattle in aviation and other economic sectors, yet even that was a stretch as it was clear that Alaska was looking to establish its own regional flight museum. This was sizable real estate to serve, international in scope. As Captain Jack Leffler explained, "We believe in moderation in all things, even excess."

A. Elliott Merrill

Born at the start of the 20th century, Seattle native Albert Elliott Merrill crossed paths with Boeing many times before joining the Seattle company in 1941. Back in 1928, for example, he had unintentionally made the first-ever landing at Boeing Field when the OX-5 engine of his American Eagle biplane overheated. Little did he suspect that he would log 20,000 more landings there in the years to come.

As a boy, Elliott was fascinated by technology. At the University of Washington in Seattle, he pursued electrical engineering while also taking advanced ROTC training. Upon graduation in 1925, he received his degree and a commission in the U.S. Army Air Service Reserve. After military flight training in Texas, he returned to Seattle to begin earning his living as a civilian flight instructor.

In 1928, while in his mid-twenties, Merrill and a partner organized the Washington Aircraft and Transport Corporation. One of Boeing Field's first occupants, this company provided Seattle with air charter and flight instruction services. It survived the Great Depression and kept Merrill busy throughout the 1930s.

Merrill also flew for fun and gained a measure of fame barnstorming the Northwest in a Boeing Model 80A airliner that had retired from service. Pilots used to joke that the greatest danger of barnstorming was starving to death, but with 18 passenger seats, the Boeing trimotor biplane eked out a profit charging thrill-seekers as little as a dollar a hop.

World War II broke out in Europe in September 1939. Elliott Merrill soon sensed that it was only a matter of time before the United States too became involved. Turning his company over to others in 1941, he signed on with Boeing as an engineering test pilot. His skills were vital to a nation frantically mobilizing and gearing up production. These pressures only accelerated after the Pearl Harbor attack on December 7, 1941, which plunged the United States into a war for which it was unprepared.

Throughout those war years, Merrill and his fellow engineering test pilots worked exhausting hours. In this capacity, he flew the Boeing XPBB-1 Sea Ranger, Douglas DB-7Bs that Boeing built under license, and Boeing 314A Clipper flying-boat airliners that the company was modifying for wartime duty. But most of his left-seat time was in the B-17 Flying Fortress and the Boeing B-29 Superfortress.

One B-17 came apart over the Olympic Peninsula. He held it steady while his crew bailed out, and barely escaped himself before it crashed. Late in the war, he flew a specially instrumented B-29 to the Pacific Theater to study operational conditions at Guam, Tinian, and Saipan. In between, he performed the first flight of the Boeing XC-97, a pressurized wartime transport incorporating B-29 technology. At the start of 1945, he and U.S. Army Air Forces Major Curtin Reinhardt set a blistering transcontinental speed record, flying the XC-97 from Seattle to Washington, D.C., in just six hours and three minutes.

In the postwar era, Merrill tested the B-50 bomber and KC-97 aerial refueling tanker. He spent 1952 through 1956 at Boeing's Wichita plant ensuring that the B-47 Stratojet fully met Strategic Air Command requirements. Back in Seattle, he finished his career helping test the new 727 jetliner in the early 1960s.

Elliott Merrill.

Merrill retired in 1964 but remained active in Northwest aviation. A founding member of the Pacific Northwest Aviation Historical Foundation, he served as its first president and chairman of the board.

Elliott Merrill died in 1992 at age 90. Intelligent, kind, unassuming, methodical, and above all professional, he left an indelible stamp on The Boeing Company, the Museum of Flight, Pacific Northwest flying, and global aviation as a whole.

Because of the Alaskan history of their 80A, it was fitting that Alaska be included in the regional focus of PNAHF.

With the formation of PNAHF, supporters in the media began to promote the idea of a flight museum. An article by Dan Coughlin in the business section of the *Seattle Post-Intelligencer* of September 26, 1965, called for action on behalf of the Foundation. He described the benefits of inspiring young people through the aviation accomplishments of the region and giving many employees of the aerospace industry a better understanding of their legacy. Coughlin mentioned that this was an idea deserving of corporate support, but reported that "[Boeing] Company sources concede they've considered the idea but have ruled it out as a corporate project. They'd rather concentrate their efforts on the manufacturing of aircraft and space vehicles—the present and the future."

Coughlin's article defined other obstacles, including the problem of cost for large-scale facilities and the need to acquire rare artifacts that were becoming extremely expensive. His article concluded that there was hope abundant and that many sectors of the community would pitch in if a good plan could be developed, and ended with "And time's a-wastin."

On October 6, 1965, Coughlin followed up with an article that included his concern about The Boeing Company's lack of enthusiasm for the project. "Done properly, a flight museum need not become a dusty collection of ho-hums from the past. It could swing. It could become a vital part of our Seattle scene, attractive to tourists and homeowners as well as to Boeing's own employees." But so far, all the Foundation had was an aircraft hulk and a few tubs of parts.

The article also quoted Kenneth R. Hopkins, director of the Washington State Capital Museum in Olympia, who said there was every reason for an aviation museum in Seattle. Hopkins also expressed his fear that the Smithsonian's plans to build a national air museum in Washington, D.C., would siphon off all the rare historic aircraft from the Northwest as well as from other parts of the country. He warned that once the Smithsonian had an object in its collection, it would be reluctant to release it even on a loan basis. Hopkins was right. A lot of competition was out there.

PNAHF held its first annual membership meeting on May 10, 1966, in The Boeing Company engineering theater in Renton. Membership had grown to 110, and with numbers affixed to member cards, Captain Jack Leffler was Number One.

The Space Program in the 1960s

At the time of the founding of PNAHF, the Mercury 7 astronauts had been selected and were well along in training and program implementation, already recognized as new American heroes and popular celebrities. Alan Shepard had thrilled the country as the first American in space with the successful flight of Mercury Freedom 7 in May 1961, followed by John Glenn's epic first orbital flight on Friendship 7 in February 1962.

The "New Nine" group of astronauts were selected and in training, followed soon by a third group of 14 to augment the full agenda of the Apollo program. More engineers and scientists were joining up with the initial test pilots to round out this prestigious corps of space travelers. Gemini 3 with Gus Grissom flew successfully in March 1965, and Apollo flights were now being scheduled.

The additional selections of astronauts beefed up the corps for ongoing Gemini programs to prepare for the Apollo moon missions. Gemini incorporated a larger capsule with two astronauts and advanced maneuvering in the space environment. In June 1965, on Gemini IV with James McDivitt, Edward White became the first American to walk in space. By August of that year, Gordon Cooper and Pete Conrad took Gemini V to the first weeklong orbit. These were big steps in the dramatic race to the Moon. The Apollo missions to follow riveted national attention and motivated so many young people to dream of aerospace careers and adventures.

The two-person Gemini spacecraft had many technological improvements based on Mercury capsule experience.

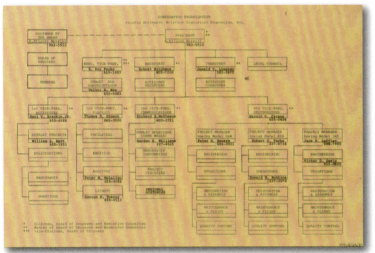

LEFFLER, Jack E. 1

MEMBERSHIP APPLICATION
PACIFIC NORTHWEST AVIATION HISTORICAL FOUNDATION, INC.
300 Rainier Avenue North   Renton, Washington 98055

Name __LEFFLER, JACK E.__   Date __Sept 14, 1965__   1-21-69

Address __P.O. Box 89166__   Phone __TR8-7950__

City __ZENITH__   State (Province) __WASH__   Zip __98188__

I wish to join the Pacific Northwest Aviation Historical Foundation, Inc., a non-profit tax-exempt educational and scientific foundation for the preservation of the history of aviation in the Pacific Northwest as a:

- ☐ Junior - Non-Voting $ 3.00
- ☒ Sustaining Member (Non-Voting)   5.00   Annual $5.00
- ☐ Voting Member   10.00
- ☐ Patron Member   Min. $25.00
- ☐ Corporate Member   Min. 50.00

Trustee – March 11, 1969

NOTE: The above dues are annual.

Top: The first PNAHF membership application, in the name of Jack E. Leffler.

Bottom: Organizational chart of PNAHF late 1965.

The nominating committee recommended 16 trustees for staggered terms of one, two, and three years of service. Eight positions of the authorized number were left open to bring on additional trustees over time. Officers were introduced, with Elliott Merrill continuing as elected president and chair of the board.

Meeting handouts included a drafted organizational chart for the Foundation in considerable detail, with portfolio areas for activities including restoration, communications, memberships, public relations, and the governance of the executive committee.

Options discussed in early meetings included the possibility of establishing a new aeronautical museum or an exhibit area within an existing institution. Local museums already had an interest in flight history. The Museum of History & Industry (MOHAI) in Seattle displayed the rare Boeing B-1 aircraft, and the Pacific Science Center promoted space themes. PNAHF also attracted interest from the State History Museum.

In a 1966 edition of the community newspaper *Highline Times*, Captain Jack portrayed the museum as a fitting memorial to Northwest aviation history featuring dramatic flying artifacts. It was evident the Captain found hunting for and acquiring aviation antiquities rewarding and adventurous. He emphasized that these objects were becoming expensive, rare, and in danger of being lost. He was quoted as saying his role was that of a collector for the new historical foundation. "What better way to measure progress than by viewing what has been produced in the past."

PNAHF was now organized, had a historic airplane, or at least a collection of its pieces, and was getting some traction. Among the members, there were enough dedicated bodies for an occasional weekend work party to parcel out tasks, and there was adequate money in the Foundation account to buy coffee. Plus, there was a growing sense of urgency that something had to be done with this precious 80A.

Early PNAHF members
were driven by the
idea of bringing
the last remaining
80A back to its
former glory.

Left: The Museum of Flight's newly restored Boeing 247 flies over Seattle's Space Needle in 2016.

Right: Boeing's 1933 advertisement celebrating the abilities of the Model 247.

One That Flies!

The Foundation was young and green, adventuresome and full of spirit. Captain Jack was on the hunt for more aircraft, and the nonprofit status gave him the license and a base for negotiations. As he flew his routes with United Airlines, he managed to recruit plane spotters throughout the aviation world.

Others in the cadre of enthusiasts searched locally, reaching out to the military and private collectors. Vintage aircraft were the priority, but engines and other smaller artifacts were also caught in the net.

Boeing archivist Brackin was in personal rescue mode for archival materials—photos, brochures, and other printed matter. He saved what he could get his hands on, accepted anything donated, and considered it all significant.

Shortly after the founding of PNAHF, Captain Jack announced that his sights were already set on another antique, but he didn't disclose its location, calling it his trade secret. Eventually, he revealed that the new prize was a rare Boeing 247D, the model commonly credited as the first modern passenger airliner.

Captain Jack had discovered the aircraft in early 1966 in California, where it had been retired in place— a polite term for abandoned—for more than five years. It was one of only four remaining. Another was destined to be suspended in the commercial aviation gallery at the new Smithsonian National Air and Space Museum, which opened a decade later.

On March 10, 1966, *Boeing News* described this aircraft's history: "Originally it was one of 69 purchased by the Pacific Air Transport division of United Airlines in 1933. This particular plane was delivered in July of that year. It was converted to a D type in 1935 and two years later

The 247 was not in great shape when Pepka and Leffler flew it from Bakersfield, California, to Renton, Washington, in 1966.

purchased by Pennsylvania-Central Airlines. Three years later it was bought by the Canadian government, used by the Royal Canadian Air Force and sold to Maritime Central Airways. Columbia Airways owned it for a while and then it turned up in California where it was used for crop dusting and cloud seeding."

This 247D had been delivered to Pacific Air Transport in 1933, and it became part of the newly formed United Airlines' fleet following the merger of Pacific, National Air Transport, Boeing Air Transport, and Varney Air Lines in 1934.

After everything had been wrung out of the airplane, it was abandoned on an airfield in Bakersfield, California, where it deteriorated. Once the queen of commercial aviation, it was in sad shape, but it was still a prize.

This aircraft was acquired in a silent bidding competition with Bill Harrah, the casino magnate, in another swashbuckling Captain Jack move. The story of this negotiation is neither entirely clear nor well documented, but it was Captain Jack who, insisting on confidentiality, led the way in the Foundation's effort to acquire the aircraft.

Harrah wanted the 247D for his collection of vintage planes and automobiles, and had reportedly offered $30,000 for it, but somehow Captain Jack and PNAHF played a winning round by using the Foundation's nonprofit status to suggest that $10,000 along with a tax advantage trumped $30,000 in cash. To complicate the transaction, PNAHF did not have the $10,000. Former museum curator Vic Seely, who was an early PNAHF member, has written that Ray Pepka came up with a check for $3,000 to cement the deal, but there is no clear trail of the full $10,000 and if it was paid and by whom. Somehow the gambit worked, and Captain Jack had again plundered successfully.

The next challenge was the ferry flight to Renton, which took some mechanical first aid, nerve, and a lot of skill. It was lucky the FAA even authorized the flight. On March 2, 1966, Captain Jack and his copilot and fellow PNAHF board member B. Ray Pepka took off in the 247 from Bakersfield. Their flight was an adventure.

According to a 1966 *Boeing News* article, "The paint was badly weathered, windows were boarded up with plywood, and much of the interior was in ragged condition. Even worse, there was concern over some of the operating parts, particularly the landing gear that had not been cycled for a number of years, so Captain Jack and co-pilot Pepka flew the aircraft at low altitude with its landing gear down."

Captain Jack and Pepka visited several airports along the route to show off the airplane. Cruising at about 120 miles an hour for the 1,000-mile journey, they stopped at Fresno and Red Bluff in California, then Medford, Eugene, and Portland in Oregon. On the final leg of their flight, they circled McChord Air Force Base, near Tacoma, and King County International Airport/Boeing Field before landing at Renton Municipal Airport.

The plane's arrival attracted media interest. Newspapers printed photographs and stories of Jack's sleight of hand in acquiring the aircraft, its return flight, and the landing at Renton. The 247D was an instant celebrity.

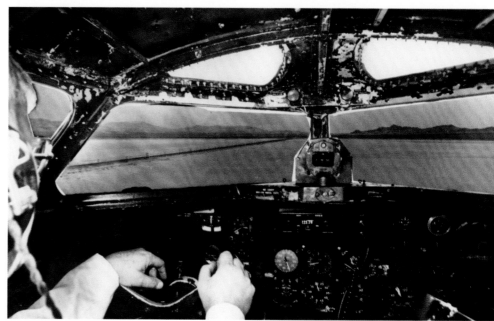

Top: Ray Pepka (left) and Jack Leffler with "one that flies!"

Bottom: Leffler flying left seat during the trip north.

Boeing 247

The museum's recently restored 247 during its final flight in 2016.

At noon on February 8, 1933, history's first modern airliner lifted off from Seattle's Boeing Field, not far from the location of today's Museum of Flight. When the Boeing 247 landed 40 minutes later, every other airliner in the world had instantly been rendered obsolete.

Boeing's new commercial offering differed profoundly from the slow, cumbersome transports then in service. Streamlined from tip to tail, the Model 247 cruised half again faster and looked futuristic with a sleek fuselage, fully cantilevered wing, and retractable main wheels. Capable of all-weather operation, it boasted de-icer boots, a gyro panel, and radio equipment permitting instrument flight and landings.

Absent were the wing struts, bracing wires, exposed engine cylinders, and other design elements typical of airliners as the 1930s began. What made this leap possible was Boeing's pioneering use of semi-monocoque construction, a transformative design paradigm that allowed airframes to be lighter, more robust, and more streamlined. The Boeing 247 was the first semi-monocoque passenger airliner ever produced. Today's jets are still built this way even though some feature composite rather than metal airframes.

The Boeing 247 carried 10 passengers and a crew of three in soundproofed, temperature-controlled comfort. Entering service in May 1933, it proved safe, versatile, easy to maneuver, and economical to operate. Launch customer United Air Lines (today United Airlines) was greatly pleased except for its marginal takeoff performance at high-elevation airfields like Cheyenne and Denver, which straddled the Rocky Mountains. The solution was to retrofit newly invented variable-pitch propellers, yielding dramatic performance improvements over fixed-pitch propellers. Boeing also upgraded the Model 247's engine cowlings and streamlined its cockpit profile, resulting in the Boeing 247D.

Having fully tied up the Boeing production line with a large initial order, United Air Lines enjoyed a competitive advantage that lasted one year. Then in May 1934, an even more advanced airliner appeared on the scene. Built by Douglas Aircraft in California, the DC-2 was larger, faster, and more comfortable. An example of the very rare DC-2 also resides in the collection at the Museum of Flight.

The Douglas DC-2 was based on the DC-1, a semi-monocoque design already under construction when the Boeing 247 first flew. The DC-1 set records and gave rise to the larger DC-2, which carried 14 passengers. Douglas engineers gave this series a higher wing loading than the Boeing, yielding a faster and smoother cruise but requiring wing flaps to keep landing speeds low, a feature the 247 didn't need. Douglas wisely set its passenger cabin well above the wing so that the wing spars wouldn't interrupt the aisle as they did on the Model 247. Although Boeing padded these obstacles well to prevent bruised shins, they weren't popular with passengers or flight attendants, who had to step over them.

Boeing's fond hopes for great commercial success foundered as the rival DC-2 sold like hotcakes to U.S. and international carriers. With better all-around performance and the potential for 40% more revenue per flight, the DC-2 was the solid hit that Boeing leaders had hoped for. Fortunately for the world, the DC-2 also gave rise to the legendary DC-3, which would put the airline industry solidly on its feet.

Boeing continued to introduce commercial transports with leading technology, but success remained elusive until the jet age. Then, starting with its famous 707, Boeing took off with a degree of success that no other jetliner manufacturer has equaled. And it all started with the humble Boeing 247.

On April 26, 2016, the Museum of Flight's resplendently restored Boeing 247—one of just four known to survive—made its final flight. Taking a break from the 787 Dreamliner program, Boeing test pilots Mike Carriker and Chad Lundy flew the vintage airliner from Everett, Washington, to the Museum of Flight, where a large and appreciative crowd welcomed it to its new permanent home.

Boeing test pilots Chad Lundy (left) and Mike Carriker bring the 247 home.

In response to a question of why he went to so much trouble to get this artifact, Captain Jack said, "I just love old airplanes, and I wanted some clean, wholesome fun." But he later quietly confessed to friends, "With the route and altitude we flew, we should have been registered with the Department of Motor Vehicles."

Never at a loss for words, Jack declared he would "advise Bill Allen at The Boeing Company to put this particular model back into production. That is how well it flies." Many years later, he would disabuse the risk angle in an interview with Sean Rossiter for the book *Legends of the Air*: "I'm not a daredevil. I've seen the elephants and heard the owls. We were careful. It was a very light airplane; I walked the throttles, and before I got to full power on the manifold gauge, we were flying. I never did make a bad landing in it, not even the first one."

On July 16, 1966, in a morning ceremony at Renton Airfield, the 247D was christened the *City of Renton* by Mrs. C. L. Egtvedt, wife of the retiring chair of The Boeing Company. This was the opening act for the Renton Aviation Festival honoring the 50th anniversary of The Boeing Company. Leffler flew the *City of Renton* in fly-bys that day and the next. Other PNAHF supporters joined in the celebration, including Clayton Scott, flying the new replica of the Boeing B&W. While Captain Jack flew around the field, his wife, Helen, drove the

Top: Evelyn Egtvedt christens the *City of Renton*.

Bottom: Elliott Merrill (front) and Clayton Scott prepare to fly the B&W replica.

PNAHF pickup in the parade, with a Boeing B-1 Flying Boat mounted atop the cab, and carrying Miss Renton candidates in the truck bed. Everyone got into the act. The young organization was up front in support of Boeing history, just as it would be 50 years later at the 100th anniversary of the company.

The Foundation's news release thanked and congratulated all the volunteers and businesses that contributed goods and services. Included was a request for funds to defray the mortgage for the purchase of the 247D and complete its restoration. It is not clear who paid off the mortgage, but it would take decades to restore the great aircraft.

The 247D flew again on August 13, 1967, at the Abbotsford, British Columbia, air show. On the return flight, piloted by PNAHF trustee Donald MacKay with Leffler as copilot, the aircraft touched down on the one wheel before it collapsed onto a grassy area just off the runway at Boeing Field. The right landing gear had dropped off upon takeoff, and the crew knew there would be difficulties in landing. This experience was reported in the news as a highly skilled touchdown, but it underscored the effort and resources required to fly historic aircraft.

The cost of maintenance and inherent risk in flying vintage aircraft represent a challenge to all flight museums. This issue often arises for debate and is not yet settled within the museum community.

Captain Jack and the board members proved their capability—their early wrangling had brought in a bigger catch at a faster pace than the group had imagined or could handle. This burden of success created a crisis that would drive the route to success.

The acquisitions of the 80A and 247D energized the group. Now the organization had two historic aircraft . . . and one that flies.

Top: The PNAHF pickup leads the parade.

Bottom: Trustee Johnny Dingle between fly-bys at the Renton Aviation Festival.

Clayton Scott taxis the Boeing B&W replica on south Lake Washington.

Jack Leffler

Captain Jack, proud of his new acquisition.

If it weren't for Captain Jack E. Leffler, the Museum of Flight would probably not have two crown jewels in its collection. In fact, the museum might not even exist.

Exuberant, talented, and seemingly interested in everything and everyone, Jack Leffler was born in 1922 in Waverly, Nebraska. As a boy, he built and flew model airplanes, read aerial adventure stories set above the trenches of World War I, and created his own pinhole camera out of an ice cream carton.

Before the United States entered World War II, Jack obtained a private pilot's license in his late teens while attending the University of Denver. After Pearl Harbor, he applied to the Aviation Cadet Program and flew with the Colorado Air National Guard to build flight time while waiting for call-up. When it came, the U.S. Army Air Forces assigned him to observation types and gliders. His determination to fly large, complex aircraft prevailed, however, and he finished up the war piloting B-24 Liberators.

United Airlines hired Leffler in 1946 and offered him a choice of bases. Jack chose Seattle and flew propeller types until transitioning to the Boeing 720—a shortened 707 for domestic routes—in the early 1960s.

Leffler loved making people laugh. A notorious practical joker, he was once grounded for two weeks after scaring passengers and a flight attendant by appearing in the cabin wearing a gorilla mask under his airline cap. Another time, he performed his airplane preflight inspection—in full view of the passenger gate windows—while wearing sunglasses and being led by a seeing-eye dog.

In the early 1960s, Leffler learned that the remains of a Boeing 80A trimotor might still exist in Anchorage, Alaska. Thrilled, he flew a charter there in 1963 and used his free time to track it down. The large biplane had been rescued from a garbage dump and was housed, disassembled, in a specially built shed. It was largely complete but in terrible shape. Leffler purchased it on the spot.

By then, he shared the growing dream of an aviation museum for Seattle. With the 80A now serving as this vision's impetus, he and others formed the Pacific Northwest Aviation Historical Foundation in 1965. The PNAHF in turn enlisted USAF, Boeing, and other support to return the 80A to Seattle for restoration.

Leffler next set his sights on acquiring the last of just four surviving Boeing 247s, the only one not yet in a museum. The plane was in Bakersfield, California, looking very sorry and no longer flightworthy. Leffler successfully negotiated its purchase, beating out other parties who had also got wind of it. Overseeing its partial restoration to flight status, he and Ray Pepka of Renton Aviation carefully flew it back to Seattle—wheels down the entire time—under a special one-time FAA ferry permit.

Finishing up a 36-year career with United, Leffler retired in 1982 as a 747 captain with 32,000 total hours in his logbook. He'd flown countless airplane types, from Mooney Mites to Grumman Mallards to civil and military gliders.

Leffler was an avid sports fan and accomplished caricaturist, but cameras and airplanes gave him the greatest joy. Combining the two, he founded a one-man sideline business called Sky Eye Aerial Photographers. From his personal Cessna 180, he captured gorgeous Northwest imagery that yielded scenic postcards plus the occasional newspaper photo. When Mount St. Helens erupted in May 1980, it was Leffler who captured the first and most dramatic aerial views.

Jack Leffler died in 1990 at age 68, leaving the Museum of Flight and Northwest flying community grieving the loss of his boundless energy, great passion, and unforgettable humor.

Left: Boeing 247 and 747 together in 1969.

Right: Clayton Scott prepares to fly the B&W during Boeing's 50th anniversary, in 1966.

CHAPTER THREE

Reach Out, Show Off, Collect

By the 50th anniversary of The Boeing Company in 1966, PNAHF had bits and pieces and a few treasures of flight history. Membership and support grew as the group reached out to entertain with the hardware in hand while trolling for more artifacts. Partners were recruited along the way, and a few began to envision this thing called a museum.

The Foundation gathered a notable coterie of history buffs, who authored a modest periodic newsletter and an annual journal of flight history. The passion of Harl Brackin continued to attract support as he continually recruited those he met inside and outside the company. Prospective volunteers assumed there was plenty of Boeing Company support and thought they were joining a juggernaut. Brackin did not disabuse them of this notion. The result was additional talent for the effort. Even though none of this brought on major fundraising experience, it did help build a cadre of supporters.

PNAHF introduced a small so-called "Engineer Office" in the surplus tower structure at Boeing Field, which on occasion was opened for tours, and was used as an interpretative center. At this point, no idea seemed bad, and the group would try anything to reach out to the community.

In the Foundation's early days, collecting vintage aircraft or archival materials was neither systematic nor particularly well organized, but the founders were on fire after the success of the 80A and 247D. The wish list grew in proportion to the membership, and available was often defined as valuable. PNAHF identified targets that ranged from wishful to questionable. For early flight aircraft, there were advocates for a Stearman C3 Mailplane, Boeing model 40-B transport, Ford Trimotor, and Pitcairn Mailwing. All of these proved elusive over the years.

Other remote possibilities were a Boeing 314 Clipper, a 307 Stratoliner, and the Boeing Dash 80, though there was to be a close encounter much later with the latter two. Others that must have seemed a stretch at the time—the B-17, B-29, and B-52—were all eventually collected. When the museum checked off a 727 on the list they had acquired the prototype. The Lockheed Electra became a prized addition 50 years later.

The collection wish list grew rapidly. Some were personal favorites of members, ranging from light aircraft and bush planes to trainers, fighters, and bombers. There was also a growing list of mail planes and commercial transports. All had some significance to aviation history in the

Top: It would take 50 years to bring this Lockheed Electra into the museum's collection.

Bottom: 80A and 247D restoration budget from the late 1960s.

Northwest and The Boeing Company. Over the years, as the mission was better defined, the selection process became more disciplined. Many on the original list survived the cut and became successful acquisitions.

During this time, PNAHF also developed a series of charts, several of which covered the restoration programs proposed for the 80A and the 247D. The plans included a proposed budget describing labor, materials, tooling, and associated costs. The estimate was $62,250 to restore the 80A to static condition and $25,900 to keep the 247D flying. These assessments displayed a naive optimism. Plans also assumed there would be big contributions of time and materials over the years. It took a while for the founders to realize what a daunting task restoring these treasures would be.

The hunt was also on for other items. Brackin was working hard to lay his hands on almost any archival materials, aerospace records, books, documents, and other collectibles that were not nailed down. Affectionate stories abound of his scooping up what he considered precious documents of company history that were scheduled for the dump or the shredder. These items, stored in members' basements or garages, would someday become part of one of the most extensive libraries and archives in the aerospace world.

PNAHF RESTORATION PROGRAMS
BOEING MODELS 80A & 247D
COST ESTIMATES

MODEL 80A

SALARY - FULL TIME MANAGER	$ 15,000.00
SALARY - FULL TIME A & E MECHANIC	10,000.00
MATERIALS FUND	
AIRCRAFT CONTROL CABLES, FTGS. $2,000.00	
ALUMINUM SHEETS 500.00	
ALUMINUM TUBING 4,000.00	
FABRIC FOR COVERING 4,000.00	
FINISHES 1,000.00	
MISCELLANEOUS HARDWARE 4,000.00	
WOOD 750.00	
MATERIALS TOTAL	$ 16,250.00
TOOLING FOR CONSTRUCTION	
(POWER TOOLS, TOOL ROOM SUPPLIES)	$ 3,000.00
ENGINES, OPERABLE	15,000.00
FENCING, SAND POINT NAVAL AIR STATION	1,500.00
TOOL HOUSE TRAILER	1,500.00
GRAND TOTAL	$ 62,250.00

MODEL 247D

ENGINE CHECK	$ 800.00
PROPELLER REPLACEMENT & REPAIR	1,500.00
STRUCTURAL REPAIR (LDG. GEAR, ETC.)	3,500.00
AIRPLANE REWIRE	5,000.00
SYSTEMS COMPONENTS REPLACEMENT	1,500.00
FURNISHINGS	5,000.00
BANK MORTGAGE BALANCE	8,600.00
TOTAL	$ 25,900.00

CHART 1.B

In spite of its small bank account, the organization—particularly Captain Jack Leffler and his colleagues—continued to look for more aircraft. Members collected quickly and desperately. It was a generalized assault. Individuals would hunt for their personal favorite, and some members looked for anything with a story or an available or neglected artifact that might need a home. They weren't always screening the targets for a particular value to regional history. In short order, other rare birds were found, acquired, and transported to Seattle.

Brackin used his company contacts on occasion to score. One of his successes was the acquisition of the surplus prototype U.S. Air Force Northrop YF-5A Freedom Fighter, one of the most valuable pieces in the Foundation's collection.

In March 1967, Brackin turned his attention to Sand Point Naval Air Station. With Commander James Korbein's support, Brackin initiated the process to acquire the 1949 Lockheed TV-1 Shooting Star and a 1953 Convair XF2Y-1 Sea Dart aircraft. With the assistance of The Boeing Company, on March 25, 1969, these were barged from Sand Point to the Renton Airport.

The TV-1 is an important aircraft in the museum collection. The Sea Dart, an unusual water takeoff jet aircraft, began a vagabond voyage that would years later

Top: The 247D during restoration.

Bottom: Assembly of the YF-5A at a temporary spot on Boeing Field.

find it leased as a roadside attraction along with the mockup of the Boeing SST in Kissimmee, Florida. The Sea Dart story underscores the challenges to a small organization trying to manage large artifacts with small budgets.

The Sea Dart was leased in the 1970s to a Florida showman who had already purchased the 1960s Boeing SST mockup. This was expected to be a short-term arrangement to make a few bucks. It didn't work. PNAHF found it difficult to collect lease payments and even harder to recover the aircraft.

Top: TV-1 being prepped for transport from Yakima to Seattle in 1968.

Bottom: This WB-47E arrived at Boeing Field in 1969, on long-term loan to the Foundation.

When the attraction in Kissimmee went bankrupt in the 1980s, the best deal for PNAHF at the time was to turn the plane over on long-term loan to the Florida Air Museum in Lakeland. Ironically, the SST mockup would years later find its way back to the Museum of Flight, while the Sea Dart still resides in Florida.

Brackin, always trolling, chatted up aviation history with an Air Force general visiting the Boeing archives, which resulted in the acquisition of the museum's B-47. On October 31, 1969, on long-term loan from the USAF, the WB-47E Stratojet Bomber touched down at Boeing Field to a gathering of members and reporters.

Nosing-around members became aware of the availability of displays at airfields and in public parks and considered these easy targets. Park queens were found across the country, from a time when the military gave them away as part of their public-relations efforts, but they had deteriorated under the exposure to weather, spray paint, and abuse, and they were often found trashed.

PNAHF took advantage of what was available and added some essential pieces, including a Grumman F9F-7, a donation from King County Parks and Recreation. Another King County Parks plane, a World War II fighter FM-2 Wildcat, had been declared a public nuisance, and it was moved from its location in White Center to the Foundation's Sand Point hangar.

The organization reached out to military units and various groups to help restore the growing fleet. With the exception of BMA's support for selected Boeing aircraft, most of these arrangements did not work out. PNAHF learned it was a lot more fun to acquire the planes than to restore and maintain them.

As word of PNAHF's collection spread, others stepped up with donations. In early 1967, Boeing supplier Pratt & Whitney donated a JT3P jet engine such as those used to power the 707 prototype. This created a long-distance partnership that would extend over several decades.

In December 1969, commercial pilot and champion unlimited hydroplane racer Mira Slovak, known as The Flying Czech, donated his Fournier RF4D, a glider powered with the engine from a Volkswagen Beetle, to the Foundation. Slovak had learned of PNAHF and its plans while serving as a personal pilot for Bill Boeing Jr.

Slovak was best known as a commercial pilot who was at the controls when he hijacked a Czechoslovakian Airlines DC-3 full of passengers and escaped from behind the Iron Curtain in 1953. In 1968, he had flown the Fournier RF4D more than 8,000 miles, from West Germany to Santa Paula, California, one of two transatlantic flights he made in the aircraft. It was the first of the museum's collection of record-setters and proved to be a favored entertainer at air shows.

In December 1971, PNAHF trucked a 1929 Curtiss Robin from Santa Paula to Seattle. The Robin was a trendy light aircraft of the late 1920s and had universal value to any flight museum. Initially on loan from two TWA pilots, the Robin was later donated to the museum and was the centerpiece of a special event that commemorated Douglas "Wrong Way" Corrigan and his flight across the Atlantic.

Top: This FM-2 served as an unofficial play structure before being rescued by PNAHF in 1969.

Bottom: "The Flying Czech," Mira Slovak, with a Bücker Jungmann circa late 1960s.

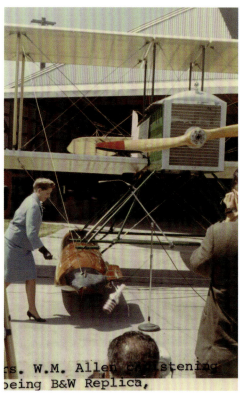

rs. W.M. Allen christening
Boeing B&W Replica,

Left: Early PNAHF members in 1971, with a 1929 Curtiss Robin that would eventually belong to the museum.

Right: Mary Ellen "Mef" Allen christens the B&W replica in 1966.

Aircraft enumerated for early acquisition included a Wright Glider replica—a must for any aviation museum; a Curtiss Pusher replica to represent the first flight in Seattle; and the 1916 Boeing B&W replica built for the company's 50th anniversary. This B&W was first loaned and later gifted to the museum. Many years later, a replica Wright Flyer would be purchased as well. In June 1971, the University of Washington donated a Wright Glider replica, which cemented a solid partnership between the museum and the UW aerospace program. In its first few years, PNAHF collected enough hardware to assure any donor that they were serious.

As aircraft and other artifacts were found and acquired, PNAHF became creative in finding ways to deliver, store, and restore them. Some were assigned to members' homes and workshops out of necessity until more permanent spaces could be arranged. Founder Kit Carson recalled storing parts of the 80A in his garage for years. The Boeing Company, along with other King County International Airport tenants, often provided trucking, towing equipment, and temporary storage. The military was the first choice for air cargo transportation of the Foundation's discoveries. All of this was done with minimal resources, relying on donated facilities and services.

The Foundation's aircraft were shuffled around from storage venue to venue as space was available. The 80A was shifted from Boeing Company Plant 2 storage to a Boeing Developmental Center, then to William E. Boeing Jr.'s hangar on Boeing Field, and from there it was moved to West Coast Airlines' hangar on the airfield. In early 1967, a 5,000-square-foot space was leased gratis in Hangar 2 at Sand Point Naval Air Station, allowing the 80A a new workspace. In 1969, The Boeing Company Contributions Committee made its first gift to the Foundation, a sum of $5,000, and the board immediately authorized $775 to construct a protective fence and some needed electrical additions in the Sand Point hangar.

During this era, PNAHF participated in air shows and aviation events in order to throttle up awareness. Working with enthusiasts and a talented pilot corps, the organization managed a lot of entertaining outreach on a modest budget. This was risky business but served as excellent promotion.

The Boeing 247D was trotted out to fly or on static display as often as possible. In November 1966, PNAHF exhibited its aircraft engine collection for viewing at Boeing Field. The presentation was part of an Aviation Historical Jamboree, which took on a carnival atmosphere with its flying demonstrations, refreshments, and auction complete with a disc jockey spinning vinyl.

PNAHF held its first dance in February 1967, featuring the Boeing Employee Band. The event raised $100 for the Foundation treasury. It took a few years for the group to realize that dances and bake sales were just for fun, not to raise serious money.

On December 29, 1967, the 247D joined the delivery of the first 727-200 aircraft to United Airlines at Boeing Field. This photographic opportunity included flight attendants in current and period attire and attracted a good deal of publicity for a delivery event.

Top: Founder "Kit" Carson and members of the Brackin family at an early PNAHF event.

Left: An early PNAHF aviation event at Paine Field.

58

Top: Delivery of the
first 727-200 in 1967.

Bottom: Ernie Gann
flies right seat in
the 247D.

The 247D flew to Paine Field in June 1970, when The Boeing Company delivered its first 747 to United Airlines. This event recognized UAL's support for the initial restoration of the 247D. At that event, author-pilot Ernest Gann flew the right seat of the 247D, with Captain Jack as the aircraft's pilot.

In July 1970, Captain Jack, Ernie Gann, and Elliott Merrill flew the 247D from Paine Field to San Francisco to join UAL's festivities in celebration of its new 747 inaugural flight to Hawaii. These appearances helped to prove the value of historical aircraft in industry promotions. Over time, this established an essential relationship between the museum and the aerospace community. The resulting support to the museum was earned.

Board members were not shy in the spotlight. In August 1970, Harold "Kit" Carson flew the 247D into a regional fly-in at Arlington, Washington, and a week earlier Captain Jack piloted the Fournier *Spirit of Santa Paula* in a fly-by over the Seafair Unlimited Hydroplane races. Well known to the speed community and general public, Mira Slovak flew various aircraft in several events on behalf of PNAHF. This small group was always ready to take to the air and perform for crowds. While busy collecting, they also managed to reach out to the community to signal that this history stuff was for real.

247D in San Francisco to celebrate UAL's inaugural 747 flight to Hawaii.

Left: The YF-5A and
Sikorsky R-4
helicopter displayed
in the Seattle Center
museum space.

Right: William E.
Boeing purchased the
1909 Heath Shipyard
on the Duwamish
River and converted
it in 1917 to Plant 1
of his new aircraft
company.

CHAPTER FOUR

Center Stage

With success in the art of collecting, PNAHF renewed interest in building a flight museum that would have the necessary space and facilities to support the mission. As early as 1966, some of PNAHF's attention was shifted to museum planning.

Harl Brackin served as the museum's dreamer-in-chief, although he professed not to have a clear vision of what a museum should or could be. He was not alone, as no one at the time had defined the qualities of a flight museum. Harl did understand that some kind of facility was required to bring all of his and others' ambitions into focus.

He talked to other corporate historians he knew, such as Harvey Lippincott at United Technologies, who also had been working on the idea of an aviation museum back east. Corporate historians were in a good position to be concerned about the loss of valuable artifacts and archival materials. They also shared the frustration of attempting to catch the attention of company leaders who were occupied with designing and building and selling new products to sustain the business. The group of historians maintained a collegial cooperation, sharing information and ideas.

Harl motored every moment he could spare in his work and personal life. He proposed various concepts and locations, seeking something that just might work to get attention and support from the industry and the community. This was a learning process, and it was also meant to develop partnerships, both of which would be key to the later successes.

He wrote that his first attempt at a museum concept was inspired by the transition of Boeing Plant 1 property on Harbor Island. As early as 1962, the Boeing facility engineer for that site told Brackin that if Building 1.05, "the Red Barn," was not moved, it would be razed.

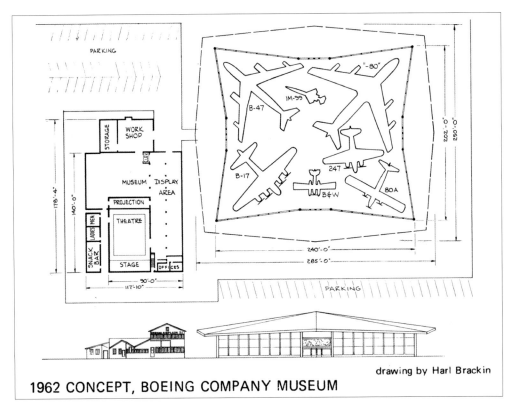

1962 CONCEPT, BOEING COMPANY MUSEUM

drawing by Harl Brackin

Harl Brackin's earliest museum concept incorporating the Red Barn.

Brackin imagined moving the building, restoring it for use as a museum, and adding an adjoining building that could hold some of the historic aircraft he and others hoped to acquire. Brackin was a trained draftsman, and he drew up plans for the site showing a large aircraft gallery structure inspired by local architect Paul Thiry's Coliseum at Seattle Center.

Brackin's drawing was probably the first idea for a flight museum complex that incorporated the Red Barn. Early documents of PNAHF contain concepts that were conjured up from time to time in response to some interest in the project. Some of the illustrations were artist's renditions from members of the board who could handle a pencil. Others, from The Boeing Company art group, were more professional. One can see the evolution of thought as they grew better and more viable.

Likely and unlikely sources presented alternative sites to the Foundation, and Brackin and PNAHF seized upon each one to gauge what kind of support it could offer. Suggested sites included properties in Renton; at Boeing Field, where the group did have some storage space; the former Boeing Scientific Research Lab property in South Park; Marymoor Park in Redmond; and Sand Point Naval Air Station. None of these locations or concepts caught on, probably because they were rudimentary, were not well architecturally delineated, and lacked a compelling case. The Foundation's process was similar to mining for gold: sluice enough gravel and one was sure to come up with a nugget.

In May 1967 at the second annual PNAHF membership meeting, a design concept was introduced for the Foundation's headquarters and an adjoining museum to be built at Renton Airport. PNAHF maintained a close relationship with this venue and received support from the City of Renton, Renton Chamber of Commerce, and Renton Airport. This concept was presented as a "Museum of Flight." Plans were drawn internally and showed a location just off Rainier Avenue North, next to the Renton Chamber office. In this conceptual structure was a call-out for a few full-size aircraft, with the 80A and the 247D as centerpieces. Most of the space was allocated to storage and restoration work.

Its narrative explained the museum as follows:

The museum, aimed at providing adequate facilities for the collection, preservation and display of aircraft and flight equipment which have played a part in the history of the Pacific Northwest, would mean that for the first time the vast region from Wyoming to Alaska would have a central collection and display point for aeronautical items. A complete library would contain archival, photographic and publications collections.

That description seems to be the sum and substance of the vision for the Museum of Flight. It was a general idea that could be located almost anywhere. Illustrations included a long rectilinear structure proposed on three levels and focused on aircraft restoration and storage, with a reference library.

Top: 1967 PNAHF design concept for a museum at Renton Airport.

In early 1967, Brackin presented ideas to the Seattle Glider Council, a group that also envisioned a museum. This museum concept was that of an "airpark," defined as a spacious property with an active airfield environment for glider flying and training, and a museum with facilities for joint use with flight-history partners and academics.

Bottom: Concept art depicting an airpark on Lake Sammamish, circa 1967.

The pitch was made to the Interagency Committee for Outdoor Recreation at the state level for the possibility of state park bond funding. The formal proposal was entitled "Lake Sammamish Airpark," located at or near an active glider field near Issaquah, Washington.

Illustrations show a small structure, a series of gallery modules with glass in the façade. The setting is bucolic, with parking for small aircraft and the suggestion of a grass runway. Jack Olson, the leader of the Glider Council, indicated that along with a state appropriation, he believed the new King County Forward Thrust bond issue might provide funding for the project. Good ideas at the time, these potential financing sources did not materialize.

In September 1967, PNAHF leadership met with Tom Ryan, assistant director of King County Parks and Recreation, to discuss a possible museum location at Marymoor Park. The group shared a concept, illustrated by artist John Amendola, with the title "Marymoor Park Aviation Museum." It was a linear structure with two stories, a large centrally located entry, and glass-fronted sections or galleries. This image was a more aesthetic treatment of the previously proposed industrial structures.

Brackin and others continued to develop the Marymoor Park proposal, and in October 1967 a formal presentation was made to the board of King County Parks and Recreation. The proposal was rejected, which ended the chase for Marymoor Park, although the idea of an airpark on the east side of Lake Washington was of interest, and they continued to seek partners and property.

illustration by John Amendola
1967 CONCEPT, MARYMOOR PARK AVIATION MUSEUM

In January 1969, Brackin and Jack Olson of the Seattle Glider Council went to Olympia to pitch the airpark concept to officials in the office of Governor Dan Evans. In May, the twosome made the same pitch to officials at Paine Field.

The Marymoor Park Aviation Museum proposal was for a large two-story building with glass-fronted galleries.

As fair turnabout, a contingent from Arlington, Washington, including the city manager and superintendent of schools, pitched their idea for an Arlington Air Park to the PNAHF board; however, no action was taken.

In August 1970, Governor Evans wrote a letter to PNAHF to endorse locating the new museum airpark at Sand Point Naval Air Station, which had been declared surplus. Its facilities were in the process of transitioning to what would be City of Seattle ownership. Early land transfers designated the property Sand Point Park, renamed Warren G. Magnuson Park in 1977. This beautiful lakefront property was considered a prime museum site.

The museum had lots of twists and turns and involved the generation of various ideas to fit the circumstances. Possibilities for land, facilities, funding, and partnerships were forming. The fact is that one kisses a lot of frogs to find a prince.

During the continued search for a suitable site, Brackin and others also became familiar with a few Seattle government officials. In September 1967, Elliott Merrill and other board members presented museum plans to Seattle Mayor James "Dorm" Braman and requested a Seattle Center space. The mayor reacted favorably. PNAHF was offered space in Building 50, a 10,500-square-foot venue with 16-foot ceilings that also housed a fire museum.

One of the exhibit buildings left vacant after the world's fair, it was a simple structure with very few amenities and had never been intended for an afterlife as a museum. Even with these restrictions and the fact that the organization had no experience in the museum business, PNAHF seized this opportunity and signed a lease agreement with the City.

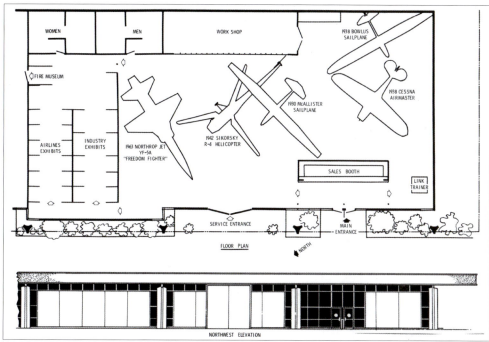

Brackin planned a layout for this first location for the Museum of Flight, and the board named him the museum's volunteer manager. Brackin carried a PNAHF business card with the title of Museum Manager, serving as the first museum director.

PNAHF welcomed the participation of other historical organizations and aviation industry companies. Brackin sent letters to potential exhibitors explaining the themes for the museum and asking for contributed services and materials that would be of interest to Seattle Center visitors. The themes read like the subject matter usually found in corporate archives. Among the display items solicited were brochures, passage tickets, clippings, historical articles, flags and banners, paintings, display models, and other such collectibles and memorabilia.

Top: Building 50 at Seattle Center, circa 1968.

Bottom: Original museum exhibit design for Seattle Center space by Harl Brackin, 1968.

The initial interpretive center concerned "the establishment, growth, and progress of local industrial and customer organizations." This prudent set of themes invited industry participation in a way that would leverage further interest in growing and building an independent museum.

On May 1, 1968, the Museum of Flight opened at Seattle Center with appropriate fanfare, some speeches, and Mayor Braman wielding the scissors to cut the ribbon. Museum operations were limited to weekends only during May and September, then functioned daily from June through August during the warmer and busier months. As the building was not heated, the museum was closed during the winter.

The museum was free but relied on a donation box. A small gift sales area was set up, with inventory relating mostly to model aircraft kits and associated items. Brackin and volunteers pitched in to clean up and prepare this awkward space as an organized interpretive center, beginning by installing a large hangar-type door to enable access for bulky artifacts. Members gathered display cases and counters and created a storage facility. Initially the museum saw few visitors, but more importantly, Seattle now had an air museum. Rent for the structure was paid to the city based on minimal gross income from the donation box and gift shop.

Loaned aircraft included a 1942 Sikorsky R-4 helicopter from Peter Jorgensen, the Northrop YF-5A Freedom Fighter, a 1930 McAllister sailplane, a 1938 Cessna Airmaster from Harold "Kit" Carson, Pete Bowers' 1938 Bowlus Baby Albatross sailplane, and the 1916 Boeing B&W replica from The Boeing Company.

An impressive collection of aircraft engines illustrated the evolution of power plants and included the first production model 502 gas turbine from The Boeing Company and an original JT3P Pratt & Whitney engine. In addition to the aviation artifacts was a library-worthy collection of photographs, magazines, books, and other materials.

Top: The doors of the Museum of Flight opened on May 1, 1968.

Bottom: Boeing's B&W replica was a prominent part of the initial exhibit.

With the full-sized aircraft and engines enhanced by various displays in cabinets, the experience was undistinguished, but it proved sufficient as an attraction. It caught on. Attendance grew to tens of thousands of visitors annually. Over time, more aircraft were moved in and interpretive components such as film clips and audio messages were added.

Suddenly challenged by the admonition to "watch what you wish for," PNAHF board members found themselves tasked with operating an interpretive attraction. Sustaining the enterprise was a big job for volunteers. However, they were learning a new business. It was small and constrained, but it was a reality-based facility and an important step in learning by doing. Exhibits at the new museum were formative but improved substantially over time.

Located near the Fun Forest amusement rides, the museum's entrance profited from foot traffic. Through the windows adjoining the entry doors, center visitors could view some of the larger artifacts, which helped to draw them into the museum. It was a limited space with formative exhibits and minimal guest comforts, but it promoted the mission. The unexpected popularity of the little attraction suggested what could be accomplished if it were appropriately expanded. Even more important, this operation provided cover for some of the artifacts, which could now generate income rather than gather dust and rust, and this first impression of a museum gave birth to the abiding mission of education.

The paid staff made $1 an hour, and because they all were devoted supporters, none considered waging a union battle. When the admission was raised to $1 for an adult, this modest attraction was able to pay its entire staff and still net a reserve. It was also a test bed for volunteers.

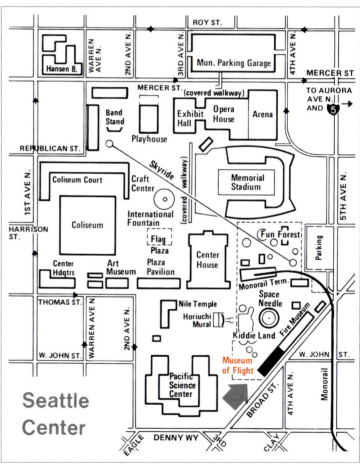

Top: A 1930 McAllister sailplane suspended from the ceiling.

1963 USAF YF-5A
NORTHROP "FREEDOM FIGHTER"

In May 1970, upon opening the doors for the third operating season, PNAHF hired Steve Brackin, son of Harl, as curator of the new museum. His duties were described as: "maintaining and operating staff to keep the museum open and clean each day, buying articles for sale, and keeping the museum books."

The Seattle Center experience was hands-on learning and instructive in real time. With this practical knowledge acquired at Seattle Center, the team was better prepared for expansion planning. Though rudimentary, it had higher visitation than the adjoining fire museum, which had a marvelous collection of artifacts. In 1978, when the Seattle Center exhibit served 100,000 visitors, it proved the acceptance of a flight museum in Seattle and a promise for what was to come. The Seattle Center site created awareness of the need for a flight museum, but it was limited in scope. Locating a suitable permanent site remained as a priority objective.

Looking back at this small group with big plans, one wonders how they managed to initiate such a bold course. Enthusiasm was high but resources limited. This was a worthy team with what at the time seemed a remote chance of any success in building a new museum. But character strengths combined with unexpected circumstances to align the stars and make this story more successful than many of the others taking place around the country. The performance at Seattle Center sent out a message.

Volunteers staff the admissions desk and small store at the Seattle Center site.

Peter M. Bowers

Pete Bowers is hailed as perhaps the finest aviation historian the United States ever produced. But impressive as this sounds, he's remembered for more than just his 26 books and 1,000 articles.

Peter M. Bowers with his Fly Baby prototype.

An active member of the Seattle flight scene for half a century, Bowers was an aeronautical engineer, aviation photographer, light-plane and glider pilot, designer and builder of homebuilt and replica aircraft, founding member of the local Experimental Aircraft Association chapter, and key contributor to what is today the Museum of Flight.

Born in San Francisco in 1918, Pete grew up during flight's exuberant golden age. It was a thrilling time, with newspapers and newsreels heralding aerial exploration, technological breakthroughs, record flights, and celebrity aviators. Shortly after Pete's ninth birthday, Charles Lindbergh flew nonstop from New York to Paris, a hugely influential event for the world and Pete.

Gravitating to airports, he took thousands of airplane pictures, on whose negative jackets he transcribed all the information he could glean about them. Like other kids, he built balsa-and-tissue model airplanes, but he didn't just play with them. Instead he tested and improved their designs with scientific rigor and then wrote articles about them, illustrating them with his own photographs. So good was his work that he soon had a national following in hobby magazines.

In his early twenties, Pete enrolled at the Boeing School of Aeronautics in Oakland, California, where he combined intensive aeronautical engineering studies with hands-on airframe and engine maintenance classes. When the United States entered World War II, he joined the U.S. Army Air Forces and served as an aviation maintenance officer in the China-Burma-India Theater. His photographic skills soon saw him drawn into intelligence duties, and by war's end his unmatched ability to categorize aircraft in aerial reconnaissance imagery led to his being placed in charge of the entire U.S. Army Aircraft Recognition Program.

Greatly impressed by Boeing airplanes during the war, Pete accepted an engineering position with the Seattle manufacturer upon leaving the service in 1947. During a Boeing career spanning 41 years, he helped preserve thousands of company photographs and much priceless history that might otherwise have been lost. Of the 26 books he authored, *Boeing Aircraft Since 1916* remains the most famous.

On his own time, Pete became a regular contributor to many aviation magazines, designed homebuilt airplanes like the prize-winning Bowers Fly Baby, helped create a Boeing B&W replica for the company's 50th-anniversary celebration in 1966, built and flew a 1912 Curtiss Pusher replica, flew and towed gliders, and so on. A Bowers Fly Baby and the B&W are today displayed in the T. A. Wilson Great Gallery.

All this activity led to close friendships with Boeing Historian Harl Brackin and countless other Seattle-area flight historians and photographers. Not surprisingly, Pete was a member of the Pacific Northwest Aviation Historical Foundation and an ardent early proponent of the Museum of Flight. It's fitting that his papers and photo collection today reside in the museum.

Pete Bowers died in 2003 at age 84, but his inspiring contributions and passion for flight live on.

Pete Bowers, shown here with a DC-3, was usually seen with his camera in hand.

Left: This 1975 museum concept features the Boeing Red Barn with several of the Foundation's artifacts.

Right: Boeing's Lake Union boathouse in 1916.

CHAPTER FIVE

Landing

The final boost to the organization came along with the challenge to save another forlorn artifact, the Red Barn. An almost forgotten remnant of early airplane building in Seattle, this structure was set aside and dismissed at its site on the Duwamish while the Port of Seattle ushered in a new era of international trade at the new Terminal 115. The Red Barn was worn out, weathered, and awaiting demolition, but it still held thousands of memories of those who had worked at Boeing Company Plant 1.

William E. Boeing did early test flying from his boathouse on Lake Union in Seattle. When he incorporated his company in 1916, he located his Pacific Aero Products plant at a boatbuilding facility he purchased for 10 dollars on the Duwamish River. This structure, now the oldest aircraft-manufacturing plant in the country, grew into what was Plant 1 of the renamed Boeing Company. The central structure, Building 1.05—the Red Barn—became an identifying symbol for this company and the aviation industry.

In the 1930s, when Plant 2 was constructed across from Boeing Field as the new, larger, and better location for the growing company, the Plant 1 property languished. The historic Red Barn was surplus to need and in danger.

In 1966, upon the 50th anniversary of The Boeing Company, this risk was accentuated. When the property was sold to the Port of Seattle for the expansion of Terminal 115 in 1970, some of the structures were demolished.

It was obvious to PNAHF leadership that this was yet another great artifact to save. As early as May 1969, Peggy Corley, King County liaison for the Federal Historic Sites Survey, began the nomination process for the historic Red Barn to status as a National Historic Place. This was an essential first move in the chess game that ensued in saving the structure and leveraging a property and new museum.

In September 1969, The Boeing Company received word that the Keeper of the National Register of Historic Places had accepted the nomination of Plant 1 building 1.05, Red Barn, as a National Historic Place. It was now a designated landmark and that made a difference.

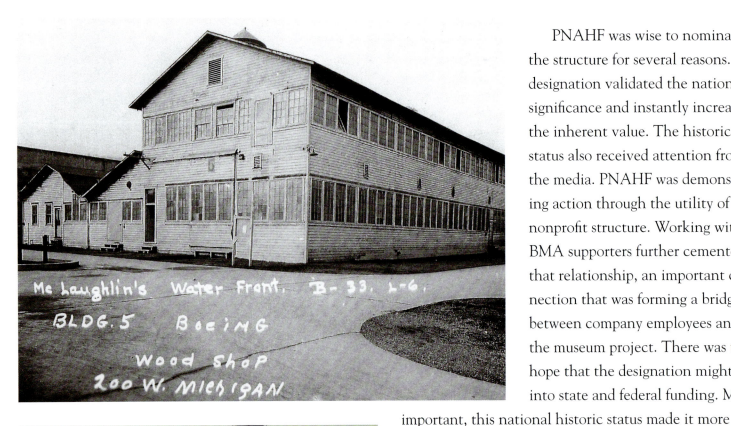

Me Laughlin's Water Front. B-33, L-6. BLDG. 5 BOEING Wood shop 200 W. MICHIGAN

Top: By the late 1930s the Red Barn had seen better days.

Bottom: Despite a new coat of paint for Boeing's 50th anniversary in 1966, the Red Barn's future was in doubt.

PNAHF was wise to nominate the structure for several reasons. The designation validated the national significance and instantly increased the inherent value. The historic status also received attention from the media. PNAHF was demonstrating action through the utility of its nonprofit structure. Working with BMA supporters further cemented that relationship, an important connection that was forming a bridge between company employees and the museum project. There was the hope that the designation might tap into state and federal funding. Most important, this national historic status made it more difficult for the Port of Seattle to destroy the building.

In February 1971, BMA members and Harl Brackin met with Port of Seattle officials to arrange for disposition of the structure. The agreement was for the Port to hold the Red Barn until PNAHF could raise funds for moving it to its museum site. However, no funds were available for such a move, and there was no museum property. It was one step at a time with a chance to save another aviation treasure.

In better times, the group could have confidently approached The Boeing Company for help. But these were the worst of times. The company was in the midst of reinventing itself, and everything was at stake.

During the late 1960s, Boeing's commercial transport sales had stalled, and many in the financial world suggested there was a better-than-even chance that the company would be another aerospace failure. Military contracts dried up, and Boeing had bet billions on the future of its first jumbo jet. The Supersonic Transport program was canceled, and the region's economy suffered accordingly. A local billboard read, "Will the last person leaving Seattle — turn out the lights."

CEO T. A. "T." Wilson and company executives were faced with difficult decisions to reverse the drain on resources and to save the company. It is written in several of the histories of

The Boeing Company that Wilson was made for such tremendous challenges, a person who could handle the most difficult decisions to reduce payroll and control expense, even though he suffered the human costs. A lesser leader and team probably would have failed, but Wilson somehow rallied the will of the company behind him and went forward to prosper in one of the world's most competitive businesses.

Harold "Hal" Haynes was Boeing's senior financial officer during this dark period and he was a committed champion and workhorse, one who watched his dark black hair turn completely white during the battle. Boeing's treasurer, J.B.L. "Jack" Pierce, was one of the key operatives under Haynes and spent much of his time with bankers, investors, and suppliers, coaxing credit line increases and buying time for the company to meet its obligations.

Pierce was young, energetic, driven, and in a position of importance as the company transitioned from the abyss back into more normal, productive, and profitable days. He was also an executive who was active in Seattle's community projects. He had a personal interest in corporate museums that was sparked by a visit to the Ford Museum and Greenfield Village, near Detroit, during a break from Boeing banking business. He was struck by the significance of recording the story of technology and business in this kind of interpretive setting. He saw the replicated Wright brothers' bicycle workshop, so essential to their work on flying. It made a lasting impression.

Pierce met Harl Brackin in the company's archives when looking for old company photographs to hang on his office walls. During his visits, he listened to Brackin's well-practiced pitch, and the museum idea, with the derelict Red Barn as an essential aerospace artifact, stuck with Pierce.

For several sensible reasons, no one in the executive offices was looking for a community project during these times, and certainly the treasurer was focused on financial issues, challenges that did not allow for much distraction. Pierce, however, was convinced that he or someone of an equivalent position in the company would have to step in to assess this museum idea.

Two views of Boeing's Plant 1 on the Duwamish Waterway: left, in 1938, when it was still producing aircraft, and right, in 1966, not long before its sale to the Port of Seattle.

T. A. Wilson was instrumental in raising funds for the Great Gallery.

On occasion, retired CEO William M. Allen, who was on the company board and had an office at head-quarters, would drop in to visit Pierce. Mr. Allen would ask about the historical photos covering the office walls, and this gave Pierce an opportunity to describe Brackin's zeal and PNAHF's plans for an aerospace history museum to the retired chief. It was neither a time to act nor to step up and build a museum, but it was an important time to listen.

In December 1974, Pierce volunteered to evaluate the feasibility of PNAHF's proposal to establish an air museum, provided that he could have at least tacit corporate indulgence. He received enough support to launch his assessment. PNAHF's enthusiasts were reas-sured that they now would have an assessment review directly from corporate offices. What might not have been anticipated was that this was not going to be a fun time, but a crunch time.

Pierce was not just energetic; he was impatient and action-oriented. His job was to find out whether there was any substance to this whole idea of a museum, and if it merited the support of The Boeing Company. His task was considerable, budget limited, and time short. As luck would have it, he was the best kind of person for the job.

For the project this was a bonus, but for those intimately involved it meant this was to become a hard-charging and demanding year. There was a lot to be done in such a short time. It was to be a sprint to decision-making. Those who worked on this task will never forget the pace.

Pierce focused on several important issues. Was there a real need for the museum? Perceived needs were not going to fly. Was there any community interest in a flight museum in this region? And if so, could public support, along with that of The Boeing Company and the industry, build and sustain the institution? Where should a flight museum be located? What was its mission?

Pierce studied the opportunities and challenges presented by the availability of the Red Barn. He also addressed the possibility of finding a suitable property for the development of such a regional museum, if it were needed and feasible, not to mention the small task of defining a flight museum. All this work was based on assumptions, foresight, and a dash of vision.

Finally, so that he could report back appropriately to corporate leaders, Pierce needed to uncover some way that this new museum organization could take on such a big job. He expected to do all this in six months. It would take several months longer.

Pierce had contacts well positioned inside The Boeing Company and within the community. He was active in various regional cultural and arts organizations and knew opinion leaders. He had experience volunteering on other community projects. Pierce organized resources within and outside the company and went to work. He did not hesitate to put pressure on those he thought could help and possessed an abiding affection for the terms "volunteer" and "pro bono." Harold Olsen, an attorney with Perkins Coie, remembered Pierce calling his office to advise him that he, Harold, would donate his legal services. And Olsen did, many times.

Pierce corralled a neighboring houseboat resident and friend, David Williams, an architect with Ibsen Nelsen and Associates, the firm that ultimately designed the first phases of the museum. Williams, Nelsen, and the firm agreed to do some pro bono work on the project, which fit well within Jack's budget. The early work of these architects was essential to the project.

Ibsen Nelsen was another character who was indispensable to the development of the museum. Nelsen was an accomplished architect and leader in the cultural community. He was admired in government and business. His professional work was elegant and added to the emerging Seattle process of civil urbanity. He brought the gift of respectability to the museum with his support and with the talent of his firm.

Boeing's treasurer, J.B.L. "Jack" Pierce, was a force in building support for the museum.

In April 1975, Pierce assigned Howard Lovering to PNAHF as his point person in coordinating the assessment to initiate planning and to assist in selling the flight museum concept to the community. Lovering, a Boeing employee working in new ventures, had reported to Pierce on other public-service projects and served as the foot messenger and planner for those tasks. Lovering received a rapid-fire explanation from Pierce regarding the museum, and later remembered it as fascinating. It appealed to his interest in developing community facilities. This was a loaned assignment of significance.

Lovering assisted Pierce with answering the abiding question: Did Seattle need an aviation museum? Today this question is rhetorical, but then the answer was not so clear.

Who would finance, build, and operate such an institution? At this time Seattle defined a large contribution as a six-digit sum. This was the major hurdle confronted for many years. Was it a valid expectation that The Boeing Company, perhaps together with the Boeing family, should preserve this industrial story, or should the community assume responsibility for creating the institution, and if the latter, to what extent? If this new museum was to be a celebration of The Boeing Company, would it not be in its interest to take on the project? This was a downtown Seattle argument that the planning team addressed.

William M. Allen

Bill Allen is hailed as one of the finest corporate leaders in history. Named president of Boeing in 1945, he led the Seattle company to unmatched success in the jet age, broadening its focus to also include missiles and spacecraft.

Born in 1900, William McPherson Allen attended the University of Montana and Harvard Law School before joining the Seattle law firm providing legal services to Boeing. The company was so impressed with him that it made him a director in 1930 and hired him away the following year to be its legal counsel.

When the United States entered World War II, Boeing contributed the war-winning B-17 Flying Fortress and B-29 Superfortress, two legendary bombers crucial to Allied success. In the midst of all this activity, Boeing President Philip Johnson died unexpectedly in 1944, and Bill Allen was named his successor the following year.

Military contracts were summarily canceled at war's end, challenging many aviation companies to survive. In contrast, Boeing prospered in the immediate postwar era thanks largely to Allen, who combined vision, piercing intelligence, business acumen, and a willingness to take calculated risks.

A good listener, Bill Allen set a tone of personal integrity and honorable dealings that inspires the company to this day. Like founder William Boeing himself, he believed the key to long-term business success was to build better products and let the world beat a path to your door. Under his leadership, the revolutionary Boeing B-47 Stratojet flew in 1947, followed by the Boeing B-52 Stratofortress in 1952.

On the commercial front, Allen gambled the company's fortunes on the Boeing 367-80 of 1954, an internally funded prototype that showed the world Boeing's vision for what a commercial jet transport should be. Never intended for production, the one-of-a-kind "Dash 80" gave rise to the similar but larger Boeing 707 jetliner of 1958. Because the Dash 80 defined the modern airliner, historians consider it the second most significant airplane in history, after the Wright 1903 Flyer itself.

The 707 established Boeing as the world's leading manufacturer of commercial jets. Under Allen's leadership, the company followed up with the 727 of 1964, the 737 of 1968, and the 747 of 1970. Famous as the world's first twin-aisle or *widebody* airliner, the 747 fundamentally redefined the air travel experience.

Allen had "bet the farm" on the 747 jumbo jet, which carried two and a half times as many passengers as any other airplane then in service. So huge a global industrial commitment was this program that Boeing very nearly failed, but the airplane was an instant hit when introduced to service by Pan American World Airways.

The time had come for Bill Allen to retire, and the 747 was his swan song. Boeing's jumbo jet had actually capped the careers of *two* illustrious aviation leaders, as it was also in part the brainchild of Pan Am founder Juan Trippe. Great friends, Allen and Trippe relinquished the presidencies of their respective companies in 1968. Bill Allen served four more years as Boeing's chairman before retiring completely. He died in 1985 at age 85, leaving a lasting imprint on Boeing and the world.

William M. Allen with models of the B&W and Dash 80 jet transport, circa late 1950s.

The counter argument was also valid. If Boeing and its employees contributed so much to history, to the economy, to taxes, then why would the local government not step up as leaders to build this historical monument? Both arguments were made, and each was useful to one side or the other.

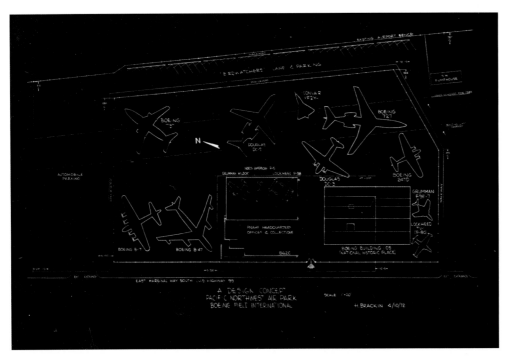

Harl Brackin's 1972 concept for a museum featuring Building 1.05 at Boeing Field.

The right answer was that there was a joint responsibility to build the museum. Correct answers are not by definition easy, and partnership would take a decade of hard work to organize.

Where should the museum be located? While working on a description of what the museum should be, the site was always front and center. It was standard for a new museum project to look for a free property somewhere that was accessible and buildable. But although property *somewhere* was thought to be a good start, that was not a good strategy because the location is vital. The planning team knew that no site was more expensive than free land in the wrong place.

Lovering's initial assignment was to evaluate the proposed museum sites, including several locations at Boeing Field, vacant acreage to the west of Seattle-Tacoma International Airport, Sand Point Naval Air Station, and The Boeing Company's South Park, along the western shoreline of the Duwamish River. Probably because the mini-museum was thriving, the Seattle Center site was also thrown into the mix, although Lovering looked upon it as a dark horse that City and Center officials did not seriously support.

Within PNAHF, there were those who saw donated sites as more attractive than those that would command a rental payment, but there were also others who found a Boeing Field site attractive. If Boeing were to get interested in this project, it would make sense for it to be near their corporate headquarters.

Boeing Field's high score went well beyond the view out the corporate window. It was based on important considerations including the property's historical significance to flight in the Northwest. The southwest portion of this airfield just happened to include the location of the region's first flight in 1910, even before the airport existed.

These and other determinants led to the selection of
Boeing Field for the Museum of Flight, but it was not a
unanimously popular decision. Some people, particularly
those in the city, wanted a more central location. Even
some board members thought there were better, more
accessible, and easier to acquire sites.

The focus was now on Boeing Field, with other
places demoted to alternates. Here was a place where the
region's history of first flight met the ongoing aerospace
story. Here again was a place where the museum had all
types of daily flight activities right in its front yard.
Here, in short, was the perfect location on which to build
a flight museum.

In June 1975, the planning team held a press
conference at the Seattle Center Museum of Flight.
With George Briggs presiding, the team presented a scale
model in concept for what was then called the Red Barn
Air Park at Boeing Field.

Briggs, who later became board chair, recalled
T. Wilson recruiting him for this presentation. Wilson
was on the board of Seafirst Bank, where Briggs was a
senior executive. During a break at a bank board meeting, Wilson asked Briggs if he would help the guys at
PNAHF with a scheduled presentation. Briggs recalled
asking, "Why would I want to do that?"

Thinking it must be imperative if Wilson had asked,
he answered his own question. This was Wilson assisting
behind the scenes, demonstrating that he had more interest in the proposed museum than he allowed publicly.

Top: One museum
location option was
the south end of
Boeing Field, visible in
this 1968 aerial photo.

Bottom: Red Barn Air
Park concept for the
Boeing Field location.

George Briggs, a respected community leader as well as a persuasive and articulate speaker, helped
dress up the presentation of the Red Barn Air Park and its public introduction. Ibsen Nelsen's
design attracted attention from the attending governmental officials, community leaders, media,
and general public. Meetings with county officials became progressive as the project gained
momentum and public enthusiasm. King County officials began to see that their support could
influence the acquisition of Boeing Field land for both the museum and other airport-related uses.

82

LEGEND
1 RED BARN
2 NEW AIR MUSEUM
3 NEW RESTAURANT
4 PARK
5 FOUNTAIN
6 GRASS VIEWING BANK
7 AIRCRAFT STATIC DISPLAY
8 PROMENADE RAMP
9 T HANGER
10 FAA BUILDING

BOEING FIELD / KING COUNTY AIRPORT – WEST SIDE DEVELOPMENT – RED BARN AIR PA
Ibsen Nelsen and Associates, Archi

Ibsen Nelsen's Air Park concept submitted as part of the West Side Development Plan.

Nelsen went on to prepare several versions and a scale model for what was called the West Side Development Plan (WSDP), which was gaining steam as a proposed package of improvements to Boeing Field. This effort provided valuable information to the airport manager and transitioned into the strategy to work with King County to expand Boeing Field and build the new museum. This integration within the airport master plan was money well invested. It introduced a cooperative team spirit between the museum and the airport that served the purpose of both. Now what had seemed impossible became just improbable. That was real progress.

In the fall of 1975 and again the following year, PNAHF chair Robert S. Mucklestone, an attorney and private pilot, Lovering, and architects Nelsen and Williams made several trips to assess museums. First on the list was the San Diego Aerospace Museum, one of the oldest in the country. Others visited were the Planes of Fame Air Museum in Chino, California; the National Air Force Museum at Wright-Patterson Air Force Base in Dayton, Ohio; the Experimental Aircraft Association Museum, then a small attraction in Hales Corners, Wisconsin; and later the new Smithsonian National Air and Space Museum (NASM) on the Mall in Washington, D.C., which opened to much excitement in 1976.

On the first trip, flying with Mucklestone, the team stopped at the Nut Tree Airport attraction in Northern California. This was a fuel and lodging stop on the way to San Diego, but it also afforded a beneficial visit that was to help shape the new museum in Seattle.

The Nut Tree owners were aviation enthusiasts with an airfield at their roadside attraction. The team traveled in a narrow-gauge train to the restaurant and lodging. In the complex was one of the finest gift shops Lovering had ever visited, featuring aviation posters, books, models, and memorabilia. The attractive design of the store and its popularity with visitors would become an influence in formulating the gift shop design for the Museum of Flight.

In Washington, D.C., the team toured the new National Air and Space Museum, meeting with senior staff, including the director, Michael Collins. PNAHF trustee Richard Taylor, then running Boeing's D.C. office, assisted with introductions. NASM was drawing huge crowds, making it the most visited institution in the country. The delegation gained fresh insights into building a flight museum. Among NASM staff were several individuals who would later serve as consultants to the MOF.

The inspiration of experiencing this great new institution was tempered by the message that high quality cost real money. The MOF team was encouraged that a flight museum could be popular, but also realized that it would cost a lot more than had been assumed.

Howard Lovering (left) and Robert S. Mucklestone during their 1975 tour of air museums.

The lesson of NASM and its unexpectedly over-the-top success was clear: people of all ages and backgrounds were curious about flight; therefore, if the collection was presented attractively and entertainingly, an aviation museum could do well. However, NASM was an elusive model, as few institutions have their resources, the collection, the unparalleled location, their funding potential, and the tremendous museum community found on the Mall.

These trips were essential to the planning and design of the MOF. Team member notes were combined with a report and slide production for presenting to the board and community groups to begin the conversation in Seattle.

The Red Barn moves slowly through the fog on the way to its new home.

The September 22, 1975, edition of the *Seattle Times* carried an editorial with the title "Aviation Park a Great Idea." It was succinct, pointing out that aviation history was essential to the growth of the region and a flight museum was timely. Ending this short comment was a plea to proceed with the project and move the Red Barn to the proposed site in time for the coming year's national bicentennial celebration. The comments from Pierce in the article spoke to the interests of the company and its involvement in the project.

Pierce and team advanced the idea that this was not just a place for aircraft and enthusiasts; it was to be an educational center with something for young and old and lots of volunteer effort. This was a very different kind of concept from what was typical at the time.

The bottom line of the PNAHF proposal was for the County to acquire the property, buy out a number of private parcels, and then to help prepare the site for the museum. This was a bold request.

With the Red Barn at risk and needing to be moved, Pierce realized that the PNAHF board needed an infusion of heavy hitters. This would take time, until the project was more promising, so Pierce and others reached out to community leaders they knew, including those not necessarily flight enthusiasts, who could give strength and credibility to governance. He restructured the board into an executive committee format, with Robert S. Mucklestone, an attorney with Perkins Coie, as PNAHF's chair.

Bill Boeing Jr. used his influence at critical points, including calling John Spellman, King County executive, to ask for assistance in saving the Red Barn. The Boeing Company and Boeing family suggested candidates from their contacts. This constituency helped to convince airport management and others that county participation with the property was appropriate, setting the museum on course.

With a tentative commitment from the County Council and the firm support of Executive Spellman, Lovering met with the director of Public Works and presented a document for signature that would allow movement of the Red Barn onto a two-acre county-owned parcel adjacent to the airfield. Lovering did his best to explain that it would just be temporary while additional adjoining properties were acquired and money raised. The director, almost without comment on the audacity of this proposal, simply signed the single-page document. With that, wheels immediately began churning to shore up and move the Red Barn before anyone had a change of mind.

Facilities specialists at The Boeing Company, supporting Jack Pierce and the team, organized the move of the Red Barn. PNAHF members and volunteers attended several work parties at the Port of Seattle property to help stabilize the structure for barge and truck conveyance. Shaughnessy Construction, an experienced structural moving and rigging company, submitted a

What better location for an air park than an active and historically significant airport?

bid for transport. This cost was met with limited PNAHF funds augmented with a $35,000 contribution from Bill Boeing Jr., who provided financial support at crucial times. The Red Barn was beloved to Boeing Jr., who as a young boy "used to go down to the barn to get balsa wood to make models." Over the years, he would step up many times on behalf of the Red Barn and museum development.

William E. Boeing Jr.

Robert Dickson, Bill Boeing Jr., and Bill Allen at the model 80A rollout in 1980.

Bill Boeing Jr., a trustee, supporter, and great friend of the Museum of Flight, grew up witnessing history unfold. Born in 1922 to aviation pioneer William Edward Boeing and his wife, Bertha, he knew his father's fledgling company from the days when its employees built fabric-covered wooden airplanes in the Red Barn. One worker there gave him a piece of balsa wood that he carved into a toy boat.

Bill was four when Charles Lindbergh flew the Atlantic in May 1927 and America became "air minded" overnight. The following year, Bill's father had a company pilot take them both up for the boy's first airplane ride. It was at brand-new Boeing Field and the large biplane was a Model 40 mail plane, the company's first commercial success.

The Roaring Twenties were in full swing. Business was booming and every day seemed to bring more headlines of aviation records and technical innovations. Then it all came to a halt when the stock market crashed in 1929, triggering the Great Depression. Boeing senior—a man of great personal integrity with a strong social conscience—covered his company's payroll out of his own savings. He also took in whatever work he could find, even having Boeing employees build household furniture. His airplane company survived when so many others didn't.

William Boeing Sr. expanded into commercial airmail and passenger operations. With aero engine pioneer Frederick Rentschler of Pratt & Whitney, he created an aviation holding company in 1929 to manage their combined interests. When in 1934 Congress passed legislation that broke up the nation's aviation conglomerates, including theirs, Boeing senior was so disgusted that he sold his aviation stocks and left the company that bore his name.

Bill Boeing Jr. grew up to pursue many interests over the course of a long and productive life. After completing his education, he founded and managed two helicopter companies, including one at Boeing Field. In the 1950s, he imported and sold Volkswagens, making them available before the German company opened dealerships in the Pacific Northwest. During that decade, he also helped promote hydroplane racing in Seattle with his much-loved racer *Miss Wahoo*.

In the 1960s, Bill's business activities included real estate development and broadcasting interests. Throughout his life, he served on many boards, but it's his philanthropy that will most be remembered. From the University of Washington to Seattle's Children's Hospital to the Museum of Flight and many other organizations, his generous support and personal involvement have quietly made a huge difference to the greater Seattle community.

Bill stepped forward years ago when help was desperately needed to save the Red Barn from destruction. He led its relocation to Boeing Field and beautiful restoration. It now serves as the nucleus of today's world-class Museum of Flight. In 2012, the museum hosted his 90th-birthday party, which was filled with friends and admirers, veteran Boeing executives, appreciative community leaders, and fellow aviation figures. He and his wife, June, were always among the most stalwart and inspirational of the museum's supporters. Fittingly, Bill was the first recipient of the museum's Red Barn Heritage Award as well as a recipient of its prestigious Pathfinder Award.

Bill Boeing Jr. died in January 2015 at age 92. Although his passing severs a treasured link with the early days of Boeing and the Seattle aviation scene, his quiet yet immense legacy lives on.

Bill Boeing Jr. is lauded at the groundbreaking ceremony for Raisbeck Aviation High School in 2011.

On December 16, 1975, Shaughnessy transported the structure to dockside, where it was hoisted upon a barge and towed from the West Waterway of the Duwamish to the Plant 2 dock at Boeing, across the street from the airfield. With aircraft and helicopters overhead, and crowds lining the route, the old building attracted attention. It was adorned with a large, bright banner with the words "Thank You America" as a tribute to the forthcoming Bicentennial. These images of the move registered regionally and nationally with the message that the Red Barn and its long history had been saved.

Once the Red Barn was ramped up onto Boeing Company property, the last leg of the delivery was the most critical. The region was in the midst of an electricians' strike, and the height of the building on the moving transporter made clearing the overhead high-tension wires a close call. One facility manager warned that the circular metal fire escape on the outside of the barn represented a potential toaster so near to the wires.

On the following morning, in a dense fog that did a lot to help control traffic on East Marginal Way, the transporter with the Red Barn crossed the arterial while electrical contractors manually raised the live power lines to provide safe clearance. With a sigh of relief, the moving crew saluted the historic structure, now positioned on the King County parcel. The move was featured as a unique Bicentennial event in *National Geographic* magazine.

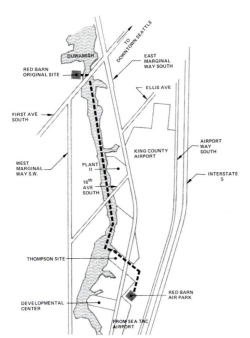

Top: The Red Barn on a barge passing under First Avenue South Bridge on its move from Plant 1 to Boeing Field.

After the December 1975 move, there was a concerted effort to prepare the Red Barn for a public showing. While this was hailed as a total volunteer effort, The Boeing Company contributed services, including assistance from experienced artists, carpenters, and painters.

PNAHF volunteers produced a new historical newsletter that was formatted to open up into a document that folded like a map. The Spring 1976 newsletter was entitled *Yesterflight*, which The Boeing Company designed and printed with articles written by museum volunteers. The first edition included a map of the move of the Red Barn and also described its significant history as well as its potential for use in a new museum.

Left: The Red Barn arrives at its temporary Boeing Field location.

Below: The first issue of *Yesterflight* was published in Spring 1976.

PNAHF members and volunteers, as well as Boeing Company artists and painters, worked after hours and on weekends to help spruce up the old structure. The growing support group, including women who were pilots or engineers at The Boeing Company, organized work teams with scout troops and their leaders. Graphic designers selected historical images and printed large photo murals to create a dramatic sense of the importance of this building.

In August 1976, with the help of The Boeing Company, the organization sponsored an open house at the Red Barn on the airport property. The team was overwhelmed by a turnout of some 20,000 people, which attested to the increased awareness of the museum's significance, and sent a message to the media, the community, and The Boeing Company.

County Executive John Spellman delivered accolades to the proposed museum. Print and broadcast media attended, and the large crowd snaked through the Red Barn to talk to the volunteers and to view the model and photographs.

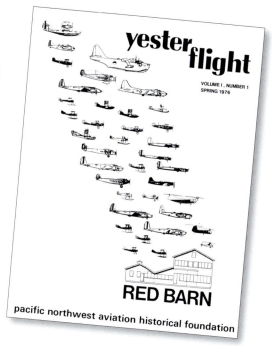

With these improvements, the Red Barn could be opened to guided tours to promote the mission and gain community support and media attention. The Allied Arts organization posted general support of the project, and as early as 1975, the *Argus* newspaper recognized the effort for outstanding achievements worthy of a "Golden Argus Man" statue. Both daily newspapers editorialized that the Red Barn restoration was worthy of community support.

An early PNAHF meeting at the newly moved Red Barn.

PNAHF's organizational structure was lean, the advancement rapid, the challenges many, and contributed services substantial. In less than one year, under the leadership and persuasion of Pierce, the team had identified the key property, finalized saving the historic Red Barn, gained permission for temporary occupancy on King County property at the airport, studied comparable museums, devised a professional museum concept that incorporated the Red Barn at the active airfield, promoted the museum, and moved the Red Barn. This had been a year of action.

After this fast-paced period, it seemed that the project had momentum and a course on autopilot. Pierce stepped back from the museum-support role to attend to his corporate duties and let PNAHF pick up the reins for implementation, though he did retain his interest and gave advice on occasion. Now reassigned to another Boeing project, Lovering was elected to the PNAHF board and continued to serve on the planning team and to represent the organization when required.

This was a good time to step back and reorganize for the next series of battles. This period lasted more than a year, and rather than rest and restoration, there was regression. Although the Seattle Center museum moved along with relative popularity, there was little other activity. After all, the museum had everything that some of the founders ever wanted. The Seattle Center operation was smoothly running with Brackin and a cadre of volunteers, and Captain Jack, more interested in locating and acquiring antique aircraft than in board work, was continuing to hunt treasures. But with the Red Barn deteriorating on its "temporary" property, another decision point was rapidly approaching.

In a series of meetings with various county officials and before the County Council, Lovering and Ibsen Nelsen continued to press the opportunity to combine the museum with the West Side Development Plan. This PNAHF pitch included a request for seven acres of land necessary to incorporate the museum. Eventually, the County Council came to a position to vote on the project at Boeing Field and its funding. Lovering and Nelsen formally presented the idea of a flight museum within the proposed airport improvements, taking questions on how it might work. It was a close vote, but favorable, and it included the appropriation by the county of both bond monies and revenue from the general fund. The important WSDP expansion, including the museum, was now approved and it appeared that PNAHF had the partner it needed.

The follow-on meetings that Lovering and Nelsen held with various county officials built productive working relationships and identified potential funding sources. Division directors made it clear that King County would not elect to run the museum even if it was at the airport and advised that they would help with the property, but no continuing support should be expected. Some argued that the land at Boeing Field was too precious to dedicate to museum use and that the costs for property acquisition were not a county priority. This opposition was further strengthened by the fact that many if not most of the tenants at the airport, as well as the airport manager, were against the museum as anything other than a minimal development.

This aerial view shows the Red Barn on blocks in its temporary location, as well as a few of the museum's aircraft.

The Red Barn at
Boeing Field before
being moved to its
permanent position.

Although not a strong union, the relationship within airport planning was the partnership needed. County Executive Spellman was a constant supporter of the museum, and with the favorable vote of the council, PNAHF was able to begin cooperative efforts with airport management to devise the new WSDP. While working to integrate the museum, PNAHF supported and promoted the related airport improvements as well.

King County Airport manager Don Smith worked to find a way to finance these needed new improvements on the west side of the airfield through several iterations. During these discussions, the museum team assisted the airport manager and his engineer in promoting the plan and finding a way to work cooperatively. Finally, Smith focused on taking back the proposed museum restaurant as a King County investment, which added estimated revenue to the package. This mix helped tip the scale in favor of financing, assessing the county project as financially feasible.

With the move of the historic Red Barn in December 1975, the early planning stage was complete. This task was a tangible accomplishment for PNAHF. The quick advancement, imaginative ideas, and bold strategies demonstrated that the project was on the move and had a chance to succeed.

Pilots avow that any landing one can walk away from is a good landing. Selecting the Boeing Field property was not just good, but a great landing. Here was a place where a flight museum could be built with flight dynamics just outside the picture window. This was hallowed ground of aviation, a place for history to meet history while history was made.

History of the Site and Airport

The history of King County International Airport/Boeing Field enriches the museum storytelling in several ways. In the late 1920s, Seattle needed an airfield to support the new but burgeoning interests of the aviation industry and flying community. After Charles Lindbergh's 1927 national tour in his historic *Spirit of St. Louis*, which did so much to promote the new age of aviation, communities across the country were identifying properties on which to build new airports. Seattle got the message and was in the hunt.

With the fledgling Boeing

Boeing Field opening celebration, 1928.

Company in a growth period and the advent of Boeing Air Transport, the region needed a municipal airport. A property near the Duwamish Waterway was ultimately selected by King County from among seven competitors. It was well situated—near to the early aviation industry and rail lines, with the waterway to the west. This decision ensured that Boeing would stay in Seattle and not move to Los Angeles.

The Duwamish property was prime for an airport and also had a fascinating history. Native Americans had long occupied this land in South Seattle, settling along what was at the time a winding Duwamish River (River of Many Colors). Early in the settlement of Seattle, the Maple family homesteaded in the general area, helping to establish what is recalled as the first public school in the region. Later, some of the Maples' land was sold to the Terry family for a summer home that they called The Onion Farm. Charles Terry, one of the pioneers who settled Alki, bought acreage in this part of the Duwamish for $1,500 in the 1850s. Terry's summer home was just a short distance from where the museum's main campus is today. The Meadows Resort and Race Track was built early in the 20th century and was a popular destination for recreation and various racing events into the 1920s.

The proposed property also had a historic aviation connection. In 1908, L. G. Mecklem flew what is recorded as Seattle's first flight, an air balloon ride from Alki's Luna Park to the Meadows Race Track. Two years later, on March 11, 1910, Charles K. Hamilton of the Curtiss Flying Team flew his Curtiss Pusher aircraft in a series of maneuvers at this same racetrack and resort. The Meadows Race Track was the southern portion of the property that would later become Boeing Field. One cannot imagine a more fortuitous combination of components for the Museum of Flight.

The King County International Airport/Boeing Field website featured the following in its airport history:

Charles Hamilton's 1910 demonstration flight at what would later become Boeing Field.

It began in 1910, when spectators turned out to see barnstormer Charles Hamilton's daring air spectacular on a piece of land known as the Meadows Racetrack—now the site of the Museum of Flight. The Meadows, which was located on the southwest corner of the present airport property, was a Coney Island type resort with horse racing, a whitewashed hotel and a boardwalk. Hamilton drew a crowd of more than 20,000 spectators that day, with adults paying $1 each to view his daredevil "dive" followed by the aircraft racing a car around the Meadows track. On the first stunt, he miscalculated the distance to the ground and crashed into a pond in the center of the race course. With some quick repairs to his plane, Hamilton flew again the next day with a young beauty contest winner at his side. The local newspaper heralded the event with the front-page headline, "Hamilton Falls, Will Fly Today!"

By 1911, the "Civic Center Project for the City of Seattle" plan, written by city planner Virgil Bogue, outlined a detailed vision that transformed the Duwamish River Valley from farmland into a developed industrial area. William E. Boeing purchased a shipyard manufacturing plant located on the west side of the Duwamish waterway, just north of where Boeing Field is now situated. In 1917, his Pacific Aero Products Company, founded in 1916, became the Boeing Airplane Company. After the United States entered World War I, the Boeing plant produced 50 Model C trainers and the airplanes were shipped across the waterway to a sandlot on the east side of the Duwamish for test flights. This testing field would eventually evolve into the property that is now King County International Airport/Boeing Field.

After Lindbergh's visit and expansion of The Boeing Company, county government went to work. King County citizens voted by an overwhelming 86% to tax themselves to acquire the land and build a modern airport. King County International Airport/Boeing Field was dedicated on July 26, 1928, at a formal gathering that included William E. Boeing and his young son Bill Boeing Jr. William Boeing Sr. is quoted as saying that day, "This is one of the most joyous days of my life."

More than 50,000 people attended the official dedication ceremonies. The following day, the Boeing Airplane Company Model 80A trimotor transport made its maiden flight from the field with Elliott Merrill at the controls. This historic aircraft was soon to replace the Model 40s on parts of the new transcontinental routes pioneered by Boeing Air Transport.

In a follow-up letter to the county commissioners, William E. Boeing Sr. wrote, "I am more deeply sensitive of the great honor because I believe that flight and air transportation are going to take a more important part in our civic and national progress than we are able to foresee today. Due to your vision of the selection of our airport, because of its location, accessibility, necessary area and physical possibilities, it is second to none in the United States." Forty-five years later, the planning team felt similarly about this location for a flight museum.

Aerial photo of Boeing Field soon after the construction of Plant 2, with the majestic Mount Rainier in the background.

Left: "Oversized Load" is an understatement as the Red Barn is moved to its permanent location.

Right: Executive Director Howard Lovering (left) with Bill Allen. PNAHF now had a stake in the ground at Boeing Field.

CHAPTER SIX

Breaking Out

In February 1977, Howard Lovering was officially hired as the museum's first paid professional director. Boeing executives, including William M. Allen, recruited Lovering and asked him to take a leave of absence from The Boeing Company to represent community-wide interests in the building of the museum. The nagging question of who would move first, The Boeing Company or the community, was implicit in this action. The company was stepping up, sending out an employee to fulfill the needed role of a full-time leader. At the same time, Lovering would leave the company payroll and work for the Foundation to organize community support. This assignment took the project to the tipping point, no turning back, and sent a challenge to both industry and the community.

After the move of the Red Barn, a lot of follow-up was necessary to identify a suitable property within the airport expansion. In turn, there was additional detailing of what the Red Barn Air Park could be as a museum. It seemed that a lot had been accomplished, and that was true, but it was wrong to think that this progress would generate momentum. It did not. The Red Barn was saved, temporarily; there was a stake in the ground at the airport location. All of this was just that, a good piece of work, but not something that was self-generating. This was progress—saving a historical structure and marrying it with an equally important property of first flight. Remarkable even. But that was where it all settled and rested and then languished. It was at this point that PNAHF faced the fact that it had to have someone working full-time to advance the project.

With Lovering coming on board as a full-time paid director, there now was someone with responsibility to represent PNAHF, someone to carry the torch, someone to blame. Surely, every project at some point needs a director, and this was the time. Lovering sometimes wondered how all of this happened so quickly, then came to understand that it was inevitable. Bob Mucklestone recalled that it was Jack Pierce who recommended  Lovering for the assignment. PNAHF leaders knew this was a critical period and that they needed someone, and reached out to The Boeing Company for support. Pierce and the company decided against providing another loaned executive to the project. It was time for PNAHF to grow up, and if someone from Boeing could help, that person should take a leave of absence and dedicate attention to the museum project.

Answering a call from Stan Little, Boeing Company vice president for Community and Industrial Relations, Lovering responded that no, he would not take the assignment. He knew that this was what amounted to a kamikaze mission and felt that he was not in a position to take that risk while responsible for a family. It is possible that other potential candidates at the company felt the same way and were much wiser in their decisions to emphatically say no to the vague and seemingly hopeless assignment. But it is probable that Lovering, who was relatively new to the company, was the only recruit targeted for the job.

Asking him to reconsider, Stan Little mentioned that Bill Allen and James Prince, senior vice president, would join the conversation. Lovering said that it was this attention from such accomplished leaders that caused him to agree to the meeting. "I asked for some considerations, was promised most of what I requested, though little of that was instituted. I said yes, I would take the temporary assignment, full of pride at being in a room full of such distinguished gentlemen. The fact that I had no contract to formally enumerate the assurances escaped me in my naiveté."

Lovering became a PNAHF employee, reporting to the board and its chair, Robert S. Mucklestone. That the Foundation had no money to pay the salary was just the first of many inconveniences of the job. He had to find a way to sustain his position, while also scrounging a modest budget for planning and operations. His new role was an awakening from a comfortable corporate position to a wild, undefined community museum building project with no budget and a limited prospect for success. It was a lonely position to occupy in the face of so many expectations.

Lovering (right) at the construction site with the two Boeing leaders who urged him to take the reins, James Prince and Bill Allen (second and third from left). Also pictured are trustees Wells McCurdy (far left) and Bob Dickson (second from right).

Though the final museum site was far from determined, it would have been hard to argue for another spot once the Red Barn was nestled next to the runway at Boeing Field.

As Lovering explained, "I was just smart enough to do the job, but not smart enough to say no to the assignment. I cannot say for sure that any of the leadership really believed the museum must be built. What I recall from that time is that there was concurrence among them that someone full-time had to push to a point of go or no-go, and I do think several expected a no-go. My year or two of work could have that result and then I would be back to a real job. In no way did I expect to be there another 15 years."

With a director in place, it was a good time to reorganize for the next series of battles. The Red Barn was deteriorating on its "temporary" property. It was entirely possible at this point that the project would fail and that the Red Barn would be moved again or destroyed. That was an option just as reasonable as thinking the community would build a new museum at the airfield. Lovering realized that he was very probably the one who would be handling a retrograde and wrapping up a project. Those were among several possibilities, and it was not clear which was most likely.

Flight Museums

A number of regional flight museums were proposed around the country in the post-WWII era, with a lot of activity in the 1960s. In almost every region, there were enthusiasts who would gather and find reasons for establishing a museum of some kind. In most cases, the interest was in preserving valuable artifacts of the past or stories of aviation history that were in danger of being lost. Aviation had made a mark. Any region could claim some greatness in its contributions to the history of flight. It might be because of a local industry, a war hero who flew, a historic airbase, the establishment of an airport, or any of the host of aviation-related subjects. The fact was and is that aviation has helped to build this country and there is interest everywhere.

Few of these proposals found easy success, although various groups did form and set out to acquire, restore, and conserve valuable artifacts, often in a clublike organizational manner. Even the Smithsonian Institution had long delays in putting together a flight museum on the mall, in spite of the fact that it had one of the most valuable and massive collections of aerospace artifacts in the world. Legendary curator Paul Garber, who had managed with his enthusiasm and doggedness to acquire the original Wright Flyer and the Lindbergh NYP *Spirit of St. Louis* aircraft among many others, was finally after many decades able to see the new National Air and Space Museum (NASM) built prominently on the National Mall. As an apt birthday present for the Bicentennial of this country in 1976, the Smithsonian NASM opened to tremendous acclaim and broad acceptance. This new treasure became one of the most popular attractions in the nation's capital. Its success was to bode well for all of those groups across the country that were hoping to build a flight museum. It was in this promising environment that the Museum of Flight and several other regional institutions came on stream.

The Wright Flyer at the National Air and Space Museum, Washington, D.C.

Initially, Lovering's office was a desk at the Seattle Center museum. In Spring 1977, with the help of board member Richard Bangert, Lovering moved into office space donated by a generous occupant in the First Interstate Bank Building in downtown Seattle. Lloyd Raab, a Seattle investor, offered this upscale rent-free space, which functioned as the museum's headquarters for more than five years. Mr. Raab was not at all related to the aerospace industry, yet gave the space as well as the contributed services of his office assistant for administrative support. The generous space, much of it unused, allowed for part-time contractual employees. These were attractive downtown work accommodations well above the capability of the formative organization. This was typical of operations in the early years, raising and using just enough of the modest donations and contributed services to advance the project.

Museum staff Marcia Johnson Witter, Howard Lovering, and Gretchen Boeing Davidson visit Bob Dickson at the 80A restoration center.

Within weeks of opening the downtown office, Lovering retained two part-time assistants. Marcia Johnson Witter was hired as an education consultant and is still active as a member and committed supporter of the museum. Gretchen Boeing Davidson, daughter of Bill Boeing Jr., signed on as a research assistant and was a key player as a member of the Boeing family, reflecting their interests and support of the institution. Lovering went into action, taking on the tasks of getting the new museum concept accepted, the land parceled, funding started, and the design delineated.

Ibsen Nelsen and Associates was put under standard American Institute of Architects contract, and compensated as funds were available. During this process, Ibsen Nelsen and his staff never faltered in their support, even when funds were low. Tony Bâby and Mel Zisfein, experienced senior managers with the National Air and Space Museum launch, were retained as consultants to help evaluate and update the museum concept plan and development schedule.

At the 1977 PNAHF annual meeting, Seafirst Bank vice president George Briggs was elected chairman and set out to streamline the board and its activities. The budget was limited, donors were reluctant to step up so early in the project, and there was some remaining tension between the corporation and the community over who should build this institution. But there was noticeable movement.

The director and staff, along with impassioned board members and volunteers, gave the project a chance to succeed through their dogged team effort. Briggs and the board, including a few of the founders, built up momentum, made promises, kept commitments, stuck out their necks, and took the steps necessary to develop essential museum resources. The project was most often broke, but it did manage to keep up the pace for the next few years with this team effort.

Lovering asked the executive committee of the board to either close the Seattle Center operation or remodel. Lovering believed that although the museum was relatively successful, it did not adequately communicate the level of quality of a new museum and could be detrimental to fundraising efforts. The board understood the issues involved, thought it would be difficult to close the exhibit, and authorized remodeling with a budget of $14,000. The Boeing Company contributed services and assigned a production designer to work on the project.

Mel Shedivy, a Boeing Company artist, soon learned the gist of his assignment when he found that the museum's "cast of thousands" was a few dedicated staff and volunteers. A jack of many trades, and gruff as a WWII Seabee can be, Mel did everything from design to electrical and mechanical upgrades to painting, even finding a way to suspend a few aircraft with cables.

From the mockup of a space shuttle nose section, Mel built a space simulator that took youngsters on a trip to Mars. The ugly industrial interior space was masked with neutral colors, black and shades of gray that allowed the colorful exhibits and displays to shine. Area and spotlighting provided drama. Audio stations enriched the displays with such voices as Charles Lindbergh and Amelia Earhart. Engines were aligned with interpretive materials to explain power-plant development. Artifacts and images were selected for specific storytelling.

The budget was just adequate to purchase building supplies, audio equipment, and surplus materials. During this remodel, Lovering suggested a simulated classroom, a landing strip where children could sit comfortably for educational presentations and hear recorded tower communications. Lovering shopped with Shedivy at discount fabric outlets for materials that were used in display cases and on artifact mounts. At one store they scored some carpet and padding, and from this Mel cut out a simulated runway with a pronounced turnaround that was used as a seating and play area for the children, replete with a painted center stripe. It was cartoonish, but a takeoff for education programs, and was soon busy with excited students.

Shedivy was characteristic of many dedicated and talented supporters from corporate positions at The Boeing Company who joined the adventure. Lovering remembered one evening at the museum toward the end of this tortuous remodel period, with Mel, whom he greatly admired, giving his exit speech: "Lovering, I should break your back for what you got me into here. This has been an awful experience and has me considering early retirement. But, looking around, I kind of like it."

The low-budget remodel was exceptional. Years later, retired in the Bay Area, Mel Shedivy again joined an MOF team as the local representative for the 1990 San Francisco exhibit "Wings over the Pacific." He remained as tough as ever.

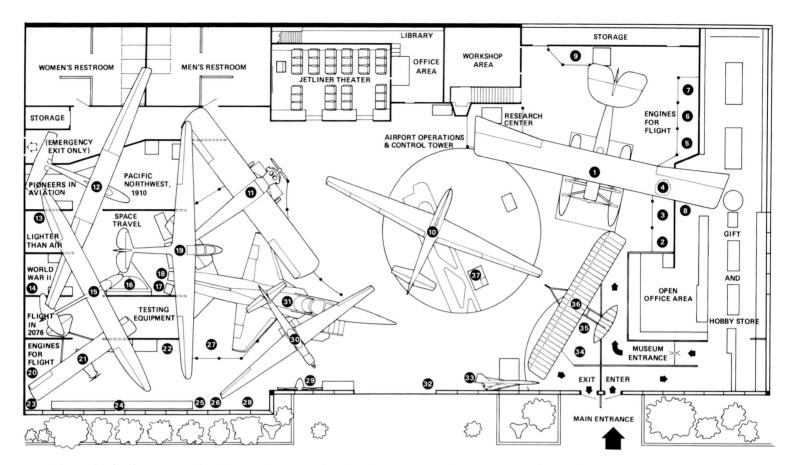

The refreshed space at the Center increased visitation, generated revenue, and gained credibility. The attraction had 100,000 visitors in 1978 and netted enough income to pay back the remodel investment and even to partially support the administrative staff. The improved gift shop also generated revenue.

With exhibits for young people and a small classroom space, volunteer Georgia Franklin began developing her successful education programs. She put in long hours and developed a reputation for her informal educational tours. Infused with energy and enthusiasm and a passion for communication, Georgia built a credible educational component. Her efforts, supported by friends and volunteers she recruited, served as ground zero for teaching in the museum setting.

Later, Georgia was added to the payroll at minimum wage. She wore many hats and was a mainstay, presenting respected programs in the museum. Franklin would receive children by the busload, sit them down on The Airfield, and run them through play activities and demonstrations, which were combined with the tour of engines and aircraft and a simulated space adventure. Over time, she organized a discovery box of items that enabled her to take the show on the road. No school was too far and no community center too small to dissuade her from loading her personal vehicle and providing outreach. In her first full year as an educator, Franklin visited 14 schools, with outreach to 500 students. In 1978, she served 66 schools, some well out of the region, and reached 3,000 students.

Enhancements to the Seattle Center space included a simulated airfield for teaching the elements of flight (center circle).

Georgia Franklin, shown here at an outreach program in 1980, promoted the museum's education mission for many years.

The improved facilities at the Center also allowed the museum to host a few educational, social, and promotional events. In December 1978, in cooperation with the American Institute of Aeronautics and Astronautics, the museum served as the venue for a joint presentation of a Diamond Jubilee lecture to salute the Wright brothers and their initial flight. The featured speaker was Jack Steiner, a design engineer, program manager, and senior vice president at The Boeing Company. This modest occasion was well received and a precursor to the many notable events to come.

Nelsen's concept drawings went everywhere with Lovering and they never failed to ramp up enthusiasm. The plan captured attention and built a fire under the Foundation's board, The Boeing Company, and the King County Council, but it would have to advance without Harl Brackin, who died in 1977. Brackin, so instrumental in conceiving this museum, was never to experience it at its final home.

In early 1979, the staff received an eviction notice for the Seattle Center exhibit, to be replaced by Fun Forest amusement improvements. They appealed to city officials, saying that the Seattle Center location was essential, and moving out would be difficult and risky to museum plans. Lovering followed Seattle Mayor Charles Royer on a radio interview and made the case that the museum served Seattle Center well and was not only in need but deserved to stay on until construction was complete in its new location at Boeing Field.

Telephone calls to the station were unanimously supportive of the museum argument. The conversation was so one-sided that Deputy Mayor Bob Royer called Lovering the next day to meet and discuss the misunderstanding. That meeting resulted in nothing more than a good dinner. Although media and public comments were supportive, all appeals failed, additional proof that the City did not consider this museum project a high priority for the community. In September 1979, the pioneer museum was shuttered.

Thirty-six years later, PNAHF founder Kit Carson bridled at the memory. "Yeah, I remember being kicked out. They threw out a museum to put in a tunnel of love." The Fun Forest improvements were made, but there never was a tunnel of love, although it did make for a good story.

Storage facilities were found for the artifacts in a warehouse near Boeing Field and shared space in a hangar at Tacoma Industrial Airport. The Boeing Company contributed transport. Someone found a forklift driver who had access to equipment that made the task a bit easier.

Red Barn woodworker and museum volunteer Carl Gustafson.

The A-team of dedicated volunteers cleaned out and loaded artifacts and materials onto the trucks. It was the usual cast of characters, including Georgia and Rod Franklin; Mike Pavone, an elderly Boeing retiree who knew his way around and had many friends; and Carl Gustafson, a lovely gentleman whose woodworking talents stemmed from early Red Barn days and who would become an interpreter in the restored Red Barn. These stalwarts and others were always present at work parties and assembled a loyal team to support the institution. It was a small group with a strong work ethic.

What was not saved and moved was available for a last-gasp event, perhaps the first museum auction. Advertised for November 16, 1979, as a "Going Out of Business Auction and Final Respects Party," the event drew a building full of members and supporters. Every surplus display case and photographic print was sold, as the crowd enjoyed one last festive evening. Lovering, as auctioneer, managed a good offer for an old upright vacuum that was promoted as having been "used to clean the Boeing 247."

As was typical, the volunteers managed a sizable task with almost no budget. Although the closure and move saddened all, it reinvigorated the planning and funding for the new museum. By June 30, 1980, Building 50 at the Center was dark and vacant. Once again, there was no flight museum in Seattle.

Although most of the artifacts and archival materials went into storage, the education program did not. All during the tedious land-acquisition process and the campaign to solicit building funds, Georgia Franklin kept the commitment to education that was the hallmark of the Museum of Flight.

Franklin continued with outreach. She developed show-and-tell items utilizing small artifacts and materials that students could touch and use. She and her colleagues developed lesson plans and assignments related to the essential learning requirements of that time. With her volunteers, she was able to continue to take aerospace education topics into the schools.

She also initiated and maintained a relationship with educational organizations in the school system and museum network. Franklin was a regular participant in meetings of the International Association of Transport Museums and became a keynote speaker on the subject of museum education programs. She was invited to travel abroad to share experiences and ideas with other institutions. Her tireless effort posted high numbers of students and schools that were served and sustained the educational mission. Lovering and the board never wavered in their commitment to education, even when staff was minimal and budgets were slim.

By 1982 the outreach program was serving students at 186 schools, a remarkable achievement that cemented the museum's commitment to education. Franklin received the regional Aerospace Educators Award from the American Society for Aerospace Education in 1980, and the Harold Wilson Memorial Award as the outstanding advocate of general aviation and aerospace in Washington State in 1982.

In July 1978, Lovering convened a planning group to reevaluate the museum concept. The team included Tony Bâby, a consultant from NASM; Fred Johnsen, an aviation historian and former editor of the organizational journal; Ibsen Nelsen, architect; Gideon Kramer, a local exhibits designer; Doug Kemper, director of the successful new Seattle Aquarium; and Georgia Franklin. The purpose was to get a variety of thoughts into the review.

Extensive thought and work went into the museum's essential theme of motivation and education through flight history and technology, and the planning team developed a comprehensive interpretive program for exhibits.

Left: Gideon Kramer Associates exhibit concept for a large, open gallery space.

Right: Trustee Bob Dickson (left) and Howard Lovering with an early model of the Great Gallery.

A week of meetings resulted in a detailed list of museum elements, including aircraft. The Ibsen Nelsen firm, with Dave Williams as design architect, revised the plan, resulting in what was to be called the Great Gallery.

This new concept was bold and unusual for any museum. It was based on the need to incorporate aircraft as large as a suspended Boeing B-17 into museum gallery exhibits. This drove the requirement for a structure with a clear span that could accommodate something of this size. It also incorporated the planning team's desire for natural light to simulate a flying environment. The design team built to the maximum square footage allowed on the property, with height and building limits for an active airfield defining the sloping roof. What appeared as something exotic and beautiful was actually a practical design solution.

This Ibsen Nelsen model of the Great Gallery illustrates the transparency that dramatizes the large open display space.

Ibsen Nelsen explained the design and engineering thought going into the expansion:

The Great Gallery roof is a space frame, a structure of connected steel members into planes, based on equilateral triangles 19'3" on each side. This clear span structure is extremely efficient and provides panel points 19'3" on center in three directions from which to suspend artifacts.

The historic B-17 weighing 40,000 pounds, for example, can be suspended at any point from the ceiling. The favorable property of the space frame is that the load is transmitted from the point of suspension evenly over the entire roof. Supporting columns are composite steel and concrete. The skin enclosing that structure is an epoxy coated aluminum framework with insulating glass.

The glass will be 65% shading, or reflective, to protect the gallery from overheating in summer, and the gallery will be air-conditioned. All air entering the museum will be filtered to protect exhibits and artifacts. The building energy design is to an energy budget developed by computer modeling.

The angle of the Great Gallery roof was designed to conform to the sloped clearance height line from the centerline of the runway of Boeing Field. The roof glass, which will reflect light, has been reviewed and cleared by the FAA and will not interfere with pilots' vision in landing or taking off.

The Red Barn's unique shape and bright color make it a standout that signifies its role as the birthplace of aviation in the Pacific Northwest. The Great Gallery walls of glass and steel will create an illusion of the illimitable space, highly appropriate for the great aircraft it will enclose. The Red Barn will be the jewel of the complex; the glass of the new gallery will reflect and enhance the color of the red barn and its white detail trim. In this way, the large new structure will highlight and not overwhelm the signature Red Barn.

Ibsen Nelsen's project architect was Lidija Gregov, who wrote in the museum newspaper about the unique features of the glass in the new gallery to explain its advanced characteristics:

Recent developments in highly sophisticated glass and glazing systems permit a high degree of transparency within realistic energy performance. The Great Gallery takes advantage of this new material. As one approaches the building, glass up to 20 feet from the ground will be clear. Above 20 feet there will be some degree of shading or reduction in transparency, the roof glazing the least transparent. The roof is composed of three layers of glass; the top has a special silver coating on its underside. The middle layer has an

The design of the Great Gallery was driven by the need to suspend large aircraft from the ceiling, such as a Boeing B-17.

additional special coating. These two layers reflect most of the sun's heat. The bottom layer consists of two sheets of heat-strengthened glass laminate with polyvinyl plastic film. This sheet reflects 99.9% of ultraviolet radiation. The system admits only 4% of solar heat and screens UV light to protect artifacts.

Sidewalls are glazed with a special "heat mirror" glass. This glass assembly is in effect triple glazing, three layers of material enclosing air spaces. The special film reflects solar heat waves while maintaining a high degree of transparency. The glass is supported by an aluminum grid into which the glass is installed with silicone sealants and gaskets. Roof members include special gutters to function in the event of seal leakage. There are provisions for absorption of vibration and thermal movement.

This was and is glass so advanced from that used in galleries and skylights at other museums.

The design drawings of the expansion and the scale model used in promotion received a lot of media coverage and caught people's imagination. The glass Great Gallery, suspended aircraft, natural lighting, and the old Red Barn—at an active airfield—were coming together. The museum organization now had a winning concept.

Design Commission Praise

In the summer of 1983, after a museum status report before the King County Design Commission, member David Scott, FAIA, credited the museum as "one of the best planned public facilities" he had ever seen. Professor Scott taught architecture at Washington State University from 1960 to 1994, serving as chair of the Department of Architecture from 1965 to 1976 and as Associate Dean of the College of Architecture and Engineering from 1976 to 1982. He received numerous honors and awards for his contributions to the design profession and public service.

Professor Scott participated in a number of museum design reviews over the years and remarked on the quality of the planning for the Museum of Flight: "From the vantage point of 34 years of study of American architecture, it is my opinion that the Museum of Flight is a very important addition to Seattle. It will be the best museum complex on the Pacific Coast. The Great Gallery is in the company of Frank Lloyd Wright's Johnson Wax Building and Philip Johnson's Crystal Cathedral. The Museum of Flight will be the new symbol for our region."

The juxtaposition of the historic Red Barn and the contemporary glass structure of the Great Gallery reflected the history of flight.

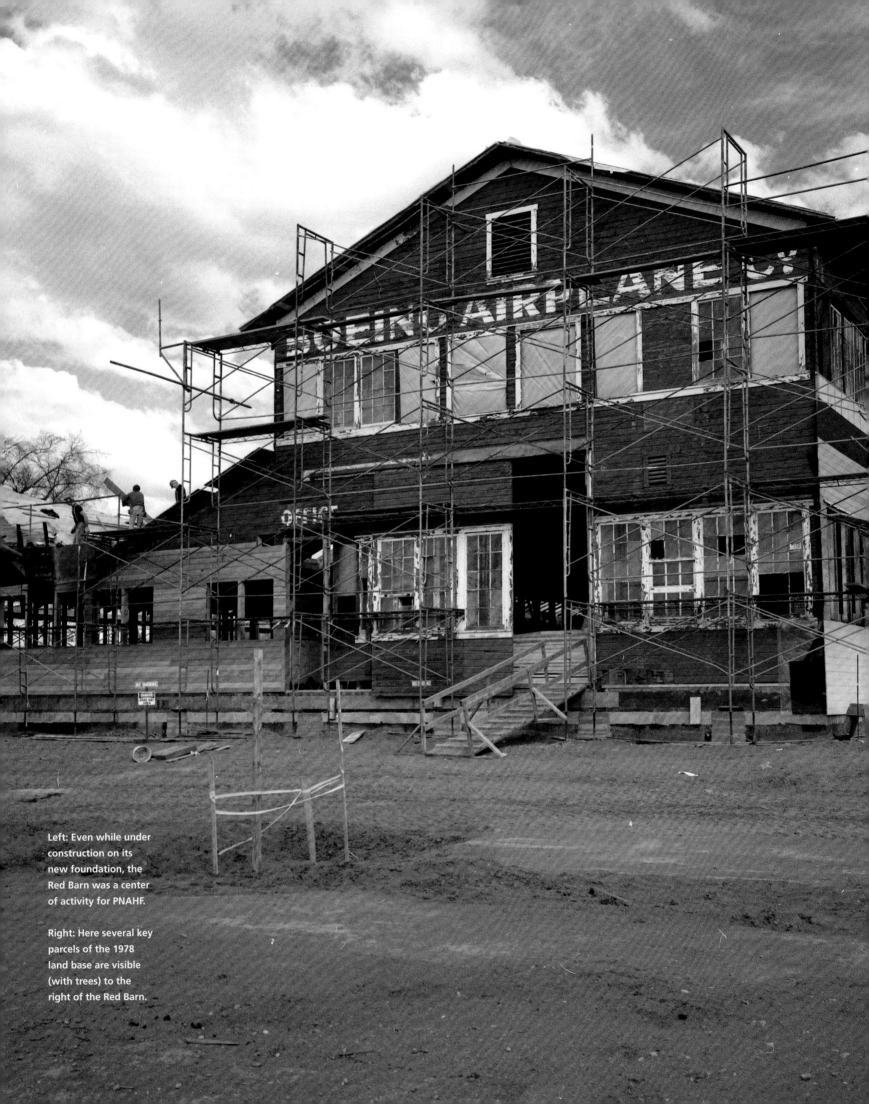

Left: Even while under construction on its new foundation, the Red Barn was a center of activity for PNAHF.

Right: Here several key parcels of the 1978 land base are visible (with trees) to the right of the Red Barn.

CHAPTER SEVEN

Breaking Ground

With progress, there were always bumps. Nothing went along smoothly. Groups within the Georgetown and South Park communities were concerned about the proposed improvements and expansion at Boeing Field, including the museum. There was a need for community outreach, and PNAHF had to help with that. A bigger problem was the burden to employ eminent domain to force acquisition of several private properties not on the market. All 13 private parcels were necessary to constitute the land base, and several owners did not want to sell or did not like the price offered. Government does not like to use eminent domain to take property, but it was the only course.

In October 1978, condemnation proceedings began in King County Superior Court for the purpose of evaluating the airport's proposal to acquire three holdout parcels. Lovering was subpoenaed to testify in the courtroom of Judge Warren Chan as a representative of the museum.

Lawyers for the landowners argued that Lovering represented an attempt to acquire privately held property for a Boeing flight museum. Lovering testified without legal representation, as the cost was not in the budget. This was a crucial time for not only the museum project but also the entire West Side Development Plan, and the deputy county prosecutor who worked the case repeatedly mentioned that he was not confident of success.

It was almost a surprise when Judge Chan ruled in favor of public use and necessity of the project and property acquisition. The judge also commented favorably on the public interest in and support for the museum, referring to it as an airport-related use that was consistent with the proposed improvements. The site was made available with this judicial determination that both the airfield improvements and the museum were needed community projects. This was a decision with language of general importance. With the favorable action, King County negotiated a price with the holdout owners, organizing the parcels into a land base for one of the major expansions of the airfield.

The County began to add these properties to those already purchased to prepare the 16-acre WSDP project and the incorporated museum. The *Seattle Times* reported on March 4, 1979, that with the council appropriation of an additional $591,000, this parceling would be completed. The total for the land acquisition was $2.7 million, a portion of which was used for demolition of structures. Second-shift factory workers bemoaned the loss of the infamous Circle Tavern on the property, to some a sorry trade-off for a museum. The article pointed out that the money for the museum's four-acre parcel came from the county general fund, while the rest of the property qualified for FAA funding and use of airport revenue.

What was not discussed was the fact that the museum property was seven acres, not four. The planners identified and designed the museum parking as an aircraft ramp with through-the-fence access so that improvement could be defined as airport related. That and the set-aside for the restaurant, then in county ownership, comprised the additional museum land for the full seven acres. The museum was off and running to prove that it was an airfield-related use that helped financially in more ways than one.

Map of the proposed museum location at Boeing Field, 1979.

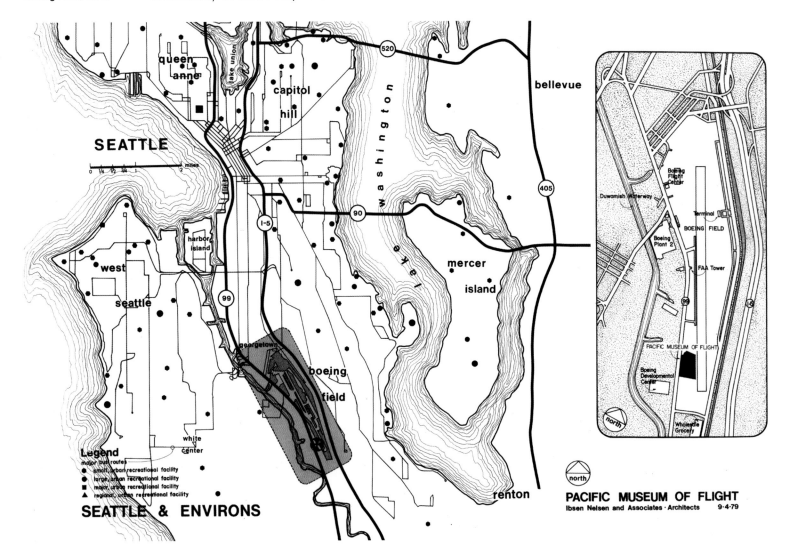

SEATTLE & ENVIRONS

PACIFIC MUSEUM OF FLIGHT
Ibsen Nelsen and Associates · Architects   9·4·79

Museum leadership celebrated this giant step. It seemed that with this prime property now in hand, all the rest would fall into place. It is true that this was an achievement, but nothing ever came easily. In fact, the euphoria of acquiring the properties was fleeting. In April 1979, the *Daily Journal of Commerce* announced that the King County Airport manager had signed a lease for a franchise restaurant on the air museum property to support the West Side Development Plan. This announcement came out of the blue. The site plan that was illustrated in the article allotted minimal space to the museum. The sketch in the newspaper featured a French farmhouse restaurant that used much of the property proposed for the museum. This theme was inconsistent with the museum design. It was a cold moment.

Lovering and Museum Chair Bangert visited County Executive John Spellman to ask him to suspend the airport restaurant lease and to put the issue of the West Side Development Plan before the King County Design Commission. Spellman agreed that this was fair. The schedule was set back, but this did provide an opportunity for the museum to join in the master planning of all the properties.

In a recent meeting to reflect on his involvement at this crucial point, Spellman said, "I did nothing special, just what I thought was right." He enjoyed discussing the museum, which he greatly respects, but he was modest in describing his contributions, though they were considerable. By remanding the airport expansion plan to the Design Commission, which the county was not legally required to do, and by making the museum a valuable component, Spellman leveled the field, and it was a significant boost.

Between May and September of 1979, monthly meetings were held with the Design Commission including Lovering, architect Nelsen, consultant Tony Bâby, and chair of the building committee, Richard Taylor, a Boeing executive who had moved back to Seattle from the D.C. area. Between these sessions, Taylor's committee met to review and prepare. As an engineer, Taylor made himself conversant with all aspects of the museum planning. Bâby, former director of exhibits at NASM, provided a national perspective.

The Commission was noticeably fascinated by the concept and how it worked within the airfield improvements. Member David Scott pronounced that the Museum of Flight would, in his opinion, be the outstanding museum on the West Coast, and one of the most important buildings in the Seattle region. This high and unexpected praise buoyed hopes for the museum team.

The Design Commission suggested that it would be well to have an integrated master plan for the West Side Development Plan, including the museum. Ibsen Nelsen was selected to prepare this document. The importance of this assignment cannot be overstated, as it allowed the Museum of Flight to be an equal partner in planning, to secure its appropriate-sized location, and to ensure that all the new investments, both airport improvements and associated development, were at a level of quality consistent with the museum.

Richard W. Taylor

Dick Taylor with a Boeing B-47, circa early 1950s.

Richard "Dick" Taylor was absolutely passionate about aviation. This lifelong dedication saw him advance the cause of flight on so many fronts that it's difficult to believe one man did it all. More than just a great engineer, he was also an accomplished test pilot and a whiz at management and policy. This combination allowed him to have an amazing impact on aerospace.

Born in Ohio in 1921, Dick grew up in neighboring Indiana and studied mechanical engineering at Purdue University, also taking Army ROTC training. Upon graduation in 1942, he received his commission and then served as an Army spotter pilot in Europe during World War II.

Less glamorous than flying fighters or bombers, piloting spotter planes like the Piper L-4—a military cousin of the venerable Cub—was in some regards more dangerous. Whether correcting the aim of field artillery, taking ground commanders aloft to reconnoiter the lines, landing in rough fields to rescue downed fliers, or racing critically injured soldiers to field hospitals, Taylor and his fellow "grasshopper" pilots took it all in stride despite being easy prey for enemy fighters and hostile fire from below.

Taylor applied to Boeing for a position as a design engineer at war's end. His engineering training, flying experience, and analytical brilliance saw him instead assigned to Flight Test. As a flight test engineer, he helped perfect the B-50 and also became the first person ever to operate the Boeing-invented refueling boom on the KB-29, Boeing's first aerial tanker.

Becoming one of the company's youngest test pilots, he progressed to flying the Boeing B-47 Stratojet, history's first large jet airplane produced with swept wings. The radically advanced B-47 flew higher and faster than other U.S. jet bombers and many jet fighters. Taylor became the Stratojet's chief project pilot, logging more than 2,000 hours at its controls.

Boundless energy and a flair for management saw him heading up Flight Test before becoming director of engineering at Boeing's Wichita Division. Returning to the Seattle area, he assumed this same role for the 737 program, before rising again to direct all product development at the Renton Division.

The 737 was a particular favorite that Taylor helped define and promote. He crisscrossed the world on sales tours, helping to establish what would become the most successful and widely sold jetliner of all time. Dick flew all Boeing jets through the 777, but it's the 737 that bears his stamp.

Taylor championed the two-pilot flight deck, demonstrating to the industry and regulatory authorities that flight engineers were no longer needed in advanced flight decks with improved systems monitoring and management. His leadership also hastened the development and use of "glass" cockpits and onboard automation.

Perhaps Taylor's greatest achievement was to lead the introduction of Extended-Range Twin-Engine Operations (ETOPS), the global regulatory framework under which two-engine jets fly long overwater routes. Four was the right number of engines in the propeller era, but data analysis showed that two fanjets yield maximum safety and reliability in the jet age. So counterintuitive was this concept that ETOPS was a tough sell, but Taylor's gifts enabled these transformative operations to begin in 1985.

Taylor's passion for flying extended into his private life. He flew aerobatics and owned and flew a wide variety of personal airplanes right up to the final years of his long life. He also set many officially certified speed records in category. His children grew up thinking that aerial family vacations were perfectly normal.

Dick Taylor died in 2015 at age 93. Larger than life, the former Boeing vice president is greatly missed and left an indelible stamp on the global industry that he loved so much.

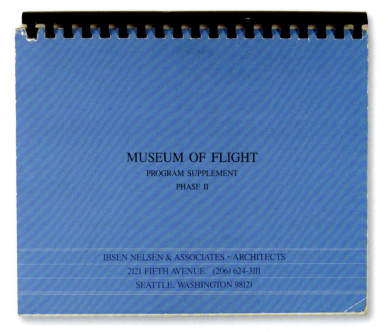

MUSEUM OF FLIGHT
PROGRAM SUPPLEMENT
PHASE II

IBSEN NELSEN & ASSOCIATES · ARCHITECTS
2121 FIFTH AVENUE (206) 624-3111
SEATTLE, WASHINGTON 98121

Top: The museum's program guide, updated as plans evolved, became known as the "Blue Book."

Bottom: Lighting was an important part of the overall campus plan.

In September 1979, Taylor held a series of meetings to refine the museum concept further. In these "building requirements" meetings, details for design, construction, and operations were reviewed with staff and consultants who had expertise in handling large artifacts, exhibits, educational and public program spaces, and visitor services. These meetings resulted in a comprehensive set of requirements, an operating model, a development budget, and a schedule that the museum board adopted. This program guide, updated as necessary as the project progressed, served as what came to be called the "Blue Book."

Ibsen Nelsen and Associates continued to refine the concept. This property had a lot of constraints to deal with, but also provided opportunities. Though there were few if any examples to study, the architects began to design a cultural and educational campus that fit within all of the restrictive guidelines.

First and foremost was the integration of the old and the new, the Red Barn and aviation gallery. This task was a delight for the design firm, working with such valuable historic resources that included the first Boeing factory building at a historic airport, with a site that had introduced heavier-than-air flight to Seattle.

In Nelsen's work for the King County Airport and its West Side Development Plan, they imagined and designed improvements in keeping with the high quality expected of the museum. The environment was industrial, but the design team developed ideas for landscaping, sidewalks, and lighting that upgraded the property and would over the years lead to major improvements in adjoining industrial properties. What was an almost barren location was beginning to evolve as an airfield oasis.

Nelsen, when asked about the wisdom of locating a cultural attraction at an industrial airport, would honestly reply that this was a magical opportunity, that the flight dynamics of an operating airport, particularly one with such a rich history, were advantages for a flight museum. In his soft but persuasive voice, he would declare the opportunity "a miracle." He meant it, and his design team acted accordingly.

After discussion and debate, the board approved a flight museum with interpretation of commercial, military, and general aviation, as well as space exploration—all four legs of the stool. This was an expanded scope, and it was not without contention.

Most board members were aircraft enthusiasts and wanted to focus on commercial transport or military aircraft. It was difficult for any regional flight museum to incorporate the space story, as laws dedicated surplus space artifacts to NASA and the Smithsonian.

Even in the early design concepts, space exhibits were an important part of how the museum was defined.

Despite the challenge involved in acquiring real stuff, the museum thought it was important to make the attempt. The discussion centered on what would attract the young, in light of the dedicated educational commitment, and that forced the decision. It was clear that young people were looking into space and not just back into aviation history.

Although there was limited opportunity for presenting large artifacts inside the Red Barn, the concept did provide for an outdoor parklike display, with exhibits inside describing the story of early flight and the building of the aviation industry. The team hoped this combination could provide a viable museum experience.

Lovering reviewed and organized the central themes necessary to tell the regional story. Professional exhibit counsel integrated interpretive requirements into building architecture, with a focus on a Phase 1 operation of the Red Barn and an administrative annex.

As with most projects, the design effort was tailored to the realities of fundraising. Requests for support were made initially to friends and family, the easiest and most interested targets, including The Boeing Company, its suppliers, and the Boeing family and business associates. Leadership was becoming pessimistic that full funding for

Early designs focused on the Red Barn and an annex.

the museum could ever be realized. Each day was a small step forward and as often one back, sometimes the slide back being much more perceptible. With all of the progress, there was yet little movement on the capital campaign. It seemed that each time the organization was poised to move, the economy was not favorable for fundraising.

Museum leadership devised a development plan allowing first for restoration of the Red Barn, securing it on the property to establish domain. The strategy was to build as much as possible, finances allowing, in a specific set of construction packages, then continue as contributions came on line. This was not a standard approach, but it also was not new or rocket science in its application. It was a way to sustain movement while the economy stabilized, always a concern in the aerospace industry, and allow time for the donor base to build. This phasing to accommodate capital budget and funding realities has persisted over the decades, refined into a continuing successful strategy at the museum.

The Pacific Museum of Flight Phase 1 Concepts and Design Guidelines for the King County Design Commission, 1979.

At its regular meeting on April 11, 1980, the King County Design Commission approved the museum's Phase 1 plan. This included restoration of the Red Barn, an entry and support building for the museum, with a gift shop, and parking for 200 cars that would ultimately be moved for the construction of the Great Gallery.

With Design Commission approval, construction and necessary funding could continue as planned. The decision as to whether or not the first phase could be opened and operational during construction of the Great Gallery was still under evaluation.

At this time the entire seven acres requested for the museum location were in the process of leasehold acquisition as part of the larger West Side Development Plan. Approximately five acres were designated for the museum complex, to include the gallery to display historic aircraft, while adjoining acreage to the north was identified for the food service and support structure, property that would eventually be taken over by the museum.

Food service was in flux, as airport management was still trying to build out this function, unsuccessful with several attempts to attract the kind of deal it wanted for a themed restaurant. The museum remained interested and was prepared to take on food service when the airport effort fell short. After several unsuccessful solicitations, the county and airport officials handed off the restaurant project to the museum. With this addition, phasing for development became even more detailed. The museum strategy was to fund a Phase 1, which included the Red Barn and support structure, then open and operate the facility for proof of concept, while organizing the larger capital campaign.

Phase 2 included the addition of the Great Gallery, exhibits of large artifacts, and ancillary site improvements. With Phase 2, the museum could function as a full-service attraction.

Phase 3 was proposed to build the attached food-service addition to the north, incorporating a café, catering kitchen, banquet room, and basement storage. With integrated food service, the museum knew it would have a much better attraction. How it would be funded was speculation at the time, but became easier with progress in construction.

As the museum concept evolved, land-scaping took on more importance in defining the overall campus.

Along the way, there would be additional steps in the process. Phase 1 used several construction packages or sub-phases in order to begin site work. Work packages were released as there was sufficient money for the contractor. This approach was risky business, but it helped show the museum's progress. As capital was raised or borrowed, additional sub-phases or construction packages were added. Thus, the initial phase was divided into five packages, with Phase 1a—the Red Barn restoration and site preparation—first under contract. The other four construction packages were to erect the new building and attendant finishes, and they came along as funding or financing was secured. This compromise between timing and budget was not as efficient as desired, but it allowed the organization to save the historic building and move ahead and leverage more credibility and support.

Funding the initial phase of the Museum of Flight was a character-building experience for the museum team. Several fundraising consulting firms conducted feasibility studies, and all had similar refrains: the campaign will be difficult, but can be successful; it will take time and more effort than anyone involved wants to believe; it will take lots of people, leaders in the community, more than anyone thinks possible to recruit; it will cost money, more than the institution has set aside. All of this proved true, though hard to swallow at the time. New projects often are tempted to believe there are funding magicians, but soon discover the realities.

Richard E. Bangert

Dick Bangert climbing into the cockpit of a Convair F-106 Delta Dart at McChord Air Force Base, 1983.

At the annual museum meeting in March 1979, Dick Bangert was elected chairman. This was a good decision from my point of view, because his office was just a few floors away, convenient for short visits and to get checks signed, although little of the latter. Thinking back to my early experiences with Dick, I have recollections of a genial, helpful person, one who was a lot friendlier than expected from someone in his position, which was chairman of First Interstate Bank at that time. Dick would always return calls and somehow find time to discuss museum matters. He took his volunteer position seriously.

It is well that we had a person such as Dick at that time. Friendly and kind, he was also stubborn. He hung in when there was abundant evidence that we should all give up. Getting the Boeing Field site from King County really wasn't looking likely, and raising the kind of money necessary to build a first-class institution was even less likely. There were several good opportunities to wrap things up, particularly when the City of Seattle abruptly asked that we move the collection out of Seattle Center. We could have just shut down the operation and there would have been no serious implications; that happens to projects.

But Dick did not fold up the tent. He stayed patient and good-natured and always had some reason for optimism, some notion that something good was about to happen. Of course, Dick would be the first to credit the others who also were helping, particularly those on the Executive Committee. That small group actually stayed together for most of the decade or more of planning and building. We always did find a way, and we were fortunate that the right thing occurred or the right person came along to help at precisely the right time. What one has to realize is that Dick Bangert was chairman through all of this period of frustration, setback, and challenge. He was the one, ultimately, who had to answer the phone to explain what it was that we planned to do next.

I recall another aspect of Dick's service with amusement. Early on he was branded as an outsider by some of the flight buffs. He was, after all, not a flier, but one of "the bankers." In fact, he was both banker and flier. Bangert was a flier all of his adult life. He had that abiding passion for flying that one finds in the best of airmen. He served his country as a young officer in the U.S. Army Air Forces, flying "The Hump" in WWII. He continued to fly light aircraft as a civilian, and regularly commuted to his San Juan Island vacation home in his private plane. He never talked about his military service or flying. True to his personality, he took it all in good spirit.

I don't know of anyone who was more popular with the museum staff than Dick. He was gregarious, always finding time to talk to ticket takers, security guards, or maintenance staff at any hour of the day, due to the fact that he kept his airplane at Boeing Field. Dick knew most everything that went on at the museum. Never did he ask for any special favors of any kind. He didn't want preferential treatment, and he always avoided taking credit.

What I'm getting around to is the point that Dick Bangert was quite a guy. He did a lot for the Museum of Flight. He certainly deserved the Chairman's award conferred upon him. He deserved even more than that. He had the respect of all of us who worked on this project over the years. One could say that he did a pretty good job for a banker.

— Howard Lovering

Dick Bangert with his wife, Betty Jane, at his retirement party in 1987.

The organization had a limited budget from membership fees that was insufficient for the tasks at hand. Early funds for planning, property acquisition, legal and financial matters, and collections were cobbled together from small gifts and grants and the generosity of the believers in the project, including Bill Boeing Jr. in concert with contributed services from a variety of sources and frequently the good graces and talent of The Boeing Company.

With progress on the land and design, the need for capital emerged as new task one. There was no longer any way to avoid what was inevitable: fund up the project or give it up. At this point, the emphasis shifted. The broad-based concept for service and the success of the formative education mission helped sell the community. There was the beginning of a persuasive case for the campaign, but it would take some time and some unexpected good fortune to find success. That would occur only after a lot of painful foot-dragging and frustration.

The museum tried many strategies for funding. One early attempt was the public-private model in which a nonprofit foundation requests government support that in turn challenges private philanthropy.

In February 1979, chair Briggs met with Jim Prince, senior vice president at The Boeing Company, to formally request a $4 million contribution. This amount was based on a 25% standard ask for a leadership gift of what was then an estimated development cost for the project of some $16 million for the Red Barn Air Park.

This formal solicitation of specific financial support was also based on the interest generated at the federal level. The state's congressional delegation was keen on the project, and U.S. Senator Warren G. Magnuson committed to searching for a federal half of the funding needed, but only if local support matched the amount. The Boeing Company was tagged for half of that local amount.

The request was forwarded to the attention of company CEO Wilson. The resulting decision was no, not at this time, but it was not a "hell no," and that was deemed to be serious progress.

In March 1979, at the Foundation's annual meeting, Richard E. Bangert was elected chair. Over the next five years under his tenure, there were continual attempts to organize a capital fund drive to build the museum. These strategies were based on private philanthropy with the awareness that there would be little money available from government. Although this is not atypical in the museum development business, it was a period of learning and frustration for all involved.

The organization hired Jay Rockey and Associates in June 1979 to establish a presence and community relations for the development drive. Rockey, a key player in the 1962 Century 21 World's Fair, was acquainted with community leaders and the process of gaining leadership. Over a period of months, Rockey counseled board and staff on building an organization with the required functional committees that were capable of a major fund drive.

In September 1979, UAL CEO Eddie Carlson, perhaps the most revered community leader, agreed to take on the museum fund drive chairmanship, doing so out of respect for Boeing Company chair Bill Allen. Carlson made it clear that he was occupied with business and many community projects and that he also saw real difficulties in raising the amount of capital needed. Once Carlson committed, George Weyerhaeuser, who also admired Bill Allen, and Bill Boeing Jr. agreed to serve as co-chairs. With these recognized leaders on board, other key individuals were recruited.

Even with this good start and the strategy set by the Rockey Company, the campaign continued to have fits and starts. A comprehensive funding brochure was published, and along with solicitation letters signed by the recognized leaders, there was an appeal to donors, including targets from a list of Boeing suppliers. The response was slow, and the realization set in that even this esteemed leadership would need a small army of support for a campaign of this size. The Phase 1 capital requirement of $5 million was a major venture.

On March 5, 1980, The Boeing Company threw a gala evening in honor of Bill Allen at Jack McGovern's Music Hall, a Spanish Baroque theater at Seventh Avenue and Olive Way in downtown Seattle. The black-tie dinner featured the Mills Brothers, dancing showgirls from Greg Thompson's Follies, and a film tribute to Bill Allen's accomplishments over the decades. Community and aerospace leaders from around the country filled the venue, including General Alexander Haig; Colonel Frank Borman, Apollo astronaut and CEO and chairman of Eastern Airlines; Juan Trippe, retired chairman and CEO of Pan American World Airways; Sir Lenox Hewitt, board chair for Qantas Airways; David Kennedy, chief executive of Aer Lingus; Donald Nyrop, retired head of Northwest Airlines; and Washington Governor Dixy Lee Ray. The dessert was individual cakes with the image of the Red Barn prominent in the frosting.

That evening, T. Wilson presented a $1 million check in honor of Bill Allen to museum foundation chairman Richard Bangert, with the comment that "the Pacific Northwest is rich in aviation history, but there is no single place where airplane buffs have easy access to historical records, equipment, and machines. That situation is going to change for the better very soon, and the group largely responsible is the Pacific Northwest Aviation Historical Foundation." Mr. Bangert, accepting the check, assured all present that this major gift "will take us a long way toward our goal of providing one of the finest historical flight museums in the country."

Boeing gives $1 million to PNAHF for Red Barn restoration

AIR MUSEUM NEWS

Pacific Museum of Flight

Vol. 1, No. 8   March, 1980   25 cents

Top: Capital campaign members (left to right): George Briggs, Howard Lovering, Dick Bangert, Bill Allen, Eddie Carlson, Bill Boeing Jr., and Wells McCurdy.

Bottom: A March 1980 issue of *Air Museum News* announced the $1 million donation from Boeing to support the Red Barn restoration effort.

The 80A was featured at Sea-Tac Airport for United Airlines' 50th anniversary of flight attendant service.

This gift at such a formal public event sent out a message that The Boeing Company was on board with the new project. The community had to see this as an endorsement, and museum leadership could proceed with some assurance. The road ahead was not easy, but it was a lot more comfortable.

At the annual membership meeting in March 1980, past chair George Briggs recapped the museum's progress. It had been an intense five years, including saving and moving the Red Barn, property negotiations with King County, victorious condemnation proceedings, concept design, the filing of impact statements, design commission reviews, and planning to ensure that the museum model was respectable and purposeful. In Briggs' words: "We have gone through a lot of red tape, a lot of promotion, a lot of hard times . . . but we are on the course we set. Are we on schedule? I'd like to quote Bob Ellis, a founder of Alaska Airlines, who said about early commercial transport in Alaska: 'We were always on schedule . . . even though we were a couple of days late.' "

Briggs announced that the planning team was scheduled to appear before the King County Design Commission on April 11 for review of the construction drawings, and with that approval the organization was ready to begin building Phase 1. He mentioned the favorable media coverage, with articles that referred to the museum as "mind-boggling" and "an opportunity to build Seattle's own Crystal Palace." Of course, there was still the matter of the dough, and it was not all in hand, but there were enough funds for a start to site work and the Red Barn restoration, an essential strategy to assure potential donors that this was a real deal.

Special events maintained the museum's exposure as a service to the community. This effort reinforced the perception that the museum was intact and awaiting its permanent facility. With small budgets, many volunteers, and lots of imagination, these events attracted an audience and captured media attention.

In May 1980, the venerable 80A aircraft was far enough along in its rebuilding program to participate in a United Airlines ceremony honoring 50 years of flight attendant service. Transported to and back from Seattle-Tacoma International Airport, the aircraft returned to Auburn, Washington, where volunteers, under the supervision of the Boeing Management Association and team leader Bob Dickson, continued their work.

PNAHF gained exposure for the museum in June 1980 by sponsoring a breakfast in downtown Seattle with General Charles "Chuck" Yeager as the guest and featured speaker. He reminisced about his experiences as a young test pilot selected to fly the Bell X-1 rocket plane in the first and successful attempt at going supersonic in 1947. Yeager, who had begun his career as a fighter pilot in the European theater of World War II, went on to a distinguished career in test flying a variety of aircraft, service in Vietnam, and then functioning as a special adviser to military aviation.

His supersonic aircraft, which he dubbed *Glamorous Glennis* after his wife, hangs in the Milestones of Flight Gallery at the National Air and Space Museum. Years later, Yeager would prove his continued support when he endorsed moving the American Fighter Aces Association headquarters to the Museum of Flight.

In July 1980, with sufficient funds for an initial construction package in hand, the Executive Committee and the board selected McCann Construction to begin work on the site.

The museum staged a colorful groundbreaking on the property on August 20. The occasion was traditional, but there was nothing usual about the format.

Among the group were a proud Bill Allen, as well as the funding co-chairs, Eddie Carlson, George Weyerhaeuser, and Bill Boeing Jr. Attention turned to the sky as local members of the U.S. Parachute Team slowly spiraled onto the property with entrenching tools in their packs. Lovering managed to get FAA approval for a short window of airspace control for the skydivers. Landing on the property, men and women outfitted in white advanced to the podium and presented their shovels to those awaiting the ritualistic business of shoveling dirt from one place to another.

The colorful festivities included a few hot-air balloons along with a band and visiting military aircraft and vehicles. It was a low-budget affair but presented enough flash and color to enliven the participants and media for this first step in building. The event signaled to the community that the museum was soon to be a reality and respected leadership was lending support to the project. Equipment moved onto the property in August 1980 and construction began.

Top left: Groundbreaking ceremonies at the Red Barn construction site, August 20, 1980.

Top right: Eddie Carlson (tan suit) oversees the first shovelfuls of (left to right) George Weyerhaeuser, Bill Boeing Jr., Carl Gustafson, and Bill Allen.

Bottom right: The signing of the museum's lease for the property on March 20, 1981.

What's in a Name?

During the first 20 years, the museum changed its name several times, which is not unusual for any cultural institution. The evolution was circular, with the original name coming back later and stronger.

When it was chartered, founders referred to the project as "the museum."

The Seattle Center attraction opened as the Museum of Flight. Years later, when making the assault on Boeing Field and a permanent site, the project was dubbed Red Barn Air Park. Saving and restoring the historic structure created a new focus.

Because of the unique airport environment, "museum" was not an operable term, as museums were not considered "airport related uses." Using the term could have created obstacles in dealing with public works officials and airport regulators at state and federal levels. Making matters worse, a few federal funding scandals caused investigations of museum appropriations, particularly the earmark variety, at several institutions around the country. Elected officials searching for federal or state funding for museum projects avoided the word *museum*. In this environment, the proposal for the Red Barn Air Park served to define an educational institution within the acceptable language of an air park, which was more easily integrated into the airport context.

As the project moved forward with land acquisition, several accomplishments changed this naming environment. The Superior Court decision in favor of King County's West Side Development Plan land acquisition mentioned the museum as an "airport related use." From that point, working with airport officials and aviation regulators, the museum managed to get this terminology entered into various documents, ultimately to include FAA administrative regulations. This fair and legal term "museum" in the context of airport uses became a Museum of Flight contribution to other flight museums attempting to locate their facilities on an airfield.

In 1982, Jay Rockey Public Relations created the museum's first professional branding, resulting in a rename to Pacific Museum of Flight, with a beautiful cloud logo. Surprisingly, this name lasted little more than a year. In interviews with prospective donors on the East Coast, the newly retained fundraising consultant learned that "Pacific" was too limiting for a national reach. Board and staff gave this quick consideration and at the end of 1982 returned to Museum of Flight. In this series of misadventures, the museum had landed back on the simplest, most recognizable, and best name for an aerospace museum.

Fearing name poaching, in August 1982 the museum applied to the U.S. Patent and Trademark Office for federal registered protection for "Red Barn" and "Museum of Flight." "Red Barn," the nickname of the

Museum of Flight
9404 East Marginal Way South
Seattle, Washington 98108
(206) 767-0118

historic Building 1.05, was granted a service mark in June 1984. "Museum of Flight" failed for a series of unfathomable reasons. Somewhere in the serpentine bureaucracy of the federal agency worked a shadowy agent dedicated to protecting the licensing world against the simple title "Museum of Flight."

Each negative ruling seemed to get less germane, and some even appeared to be outright contradictions. One ruling said something to the effect that the title was too "generic, not specific" for registration. When staff appealed with evidence of a "Museum of Lasers" and "Museum of Mass Transportation," the unassailable retort was that "they shouldn't have been registered." But the staff kept plugging, kept the paper flowing, enjoying the humor of the exchange. In August 1986 the shadowy agent gave up and let "Museum of Flight" join "Red Barn" as registered marks. It was well worth the effort.

But wait, there's more! The naming game crops up like jack when the box is opened, springing out from time to time for serious consideration or distraction. Finding the perfect name is quite a temptation, and whenever there is growth or change, one should expect the name game to resurface.

In 2003, museum leadership suggested a search to find a name worthy of the considerable expansion under way and to recognize the prominent position the museum had earned. "National Museum of Flight" was recommended, communicating more stature with broader service. Other options were placed into consideration. Bruce McCaw, Chair at the time, established an ad hoc committee to evaluate the issue, resulting in a substantial report and spirited discussion. The thorough review included a survey of other museums and their experiences with naming, while seeking opinions of the regional hospitality industry. The decision was to stay with Museum of Flight. Certainly this matter will arise again, and the simple and generic Museum of Flight will withstand another change, proving again that it is the museum that makes the name, not the name that makes the museum.

MUSEUM OF FLIGHT

Left: The Curtiss JN-4D
"Jenny" was one of
the main attractions
in the newly opened
Red Barn.

Right: Aerial view of
Red Barn restoration,
March 1981.

CHAPTER EIGHT

Barn Dance

While the first phase was under construction, the museum kept up steam with events to promote the project and its outlying property. The idea was to attract people to the neighborhood and introduce the fact that this could be a banner place to visit and have fun. It was the kind of work that carved out a niche for both the airport and the museum.

One novel and popular event was Boomerang Day. On October 3, 1981, local members of the American Boomerang Team demonstrated this sport in workshops explaining the history of "the thinking man's Frisbee" and its aerodynamic principles. These events also served as a promotion for the American Boomerang Team and their trip to Australia to challenge the current world champion team. Young and old alike purchased their own 'rang and took lessons in throwing. By the end of the day, many were proficient in getting it to return time and again. Lovering said he "was taken with the enthusiasm for the boomerang and began to imagine a variety of flying themes that could be introduced."

When night fell on Boom Day, the still-unrestored Red Barn hosted a retirement party for Frank "Sam" Houston, a captain with Northwest Airlines. Some 400 friends and associates came to honor Houston, with noted writer Ernie Gann and others paying tribute to this experienced military and commercial pilot. Frank and wife Betty were supporters of the museum and would not have the reception anywhere else. This was proof that even under construction, the museum was an appealing venue for social events.

The organization also focused on building the membership base for future funding support. The group grew from 100 members in 1977 to 2,000 by 1978, with incremental increases over the next few years. A successful Boeing employee solicitation in 1981 jetted membership to 10,000, one of the largest in the Northwest. Much of this growth was inherently artificial, as annual membership status was granted to capital gift donors, but it was critical to the perception of momentum. The museum retained a portion of its membership from year to year, building a powerful resource for sustainability.

What was particularly significant about the employee solicitation was not just the new bump in membership, but also the unexpectedly enthusiastic and generous response to the request for funding. It could not have come at a better time. Lovering sensed there was a lot of employee interest stemming from the many presentations he had made to BMA and other employee meetings, some with as many as a thousand in attendance. He asked Boeing Vice President Stan Little for permission to make a solicitation, and after several delays, it was finally approved. But there were some specific guidelines.

Little explained that CEO Wilson could not sign the solicitation letter because, at the time, he was after employee support for the proposed Vietnam Veterans Memorial, for which he served as national co-chair with Lee Iacocca. The museum had hoped for Wilson's signature but understood and also realized that to have the opportunity for an internal solicitation that allowed pledges through payroll deduction was a windfall. Finally given authority, the museum launched its solicitation with a letter signed by Bangert and Lovering covering a promotional brochure of plans; graphics and internal mailing were all donated by the company.

The Pathfinder Awards initiated a long-standing annual tribute to leaders in aviation.

It was a modest solicitation that resulted almost immediately in a flood of responses, thousands over a month. The lid was blown off with pledges amounting to $970,000, to be collected over a three-year period, with a lot of the cash up front. At this point, there was no doubt how the rank and file felt in joining management. Some consider this one of the most important moments in museum history, and that argument has merit. Later, CEO Wilson would repeat that this show of support confirmed his decision that the museum was worthy of total company support. In Autumn 1981, The Boeing Company corporate donated another $500,000 for this first phase to restore the Red Barn. The early bottom-up strategy of museum founders continued to work.

The Museum of Flight initiated the annual Pathfinder Awards in 1982 as a hall of fame. It was and is a joint venture of the museum and the regional chapter of the American Institute of Aeronautics and Astronautics. The idea, spawned even before there was a complete first phase of the museum, was to recognize leaders in the various tracks of aerospace, flying, manufacturing, engineering, operations, education, and an at-large category. In that first year,

William Boeing Sr. was inducted in the manufacturing category, former Boeing CEO Clairmont L. Egtvedt at-large, Thomas F. Hamilton and Louis S. Marsh in engineering, famous test pilot Leslie R. Tower and trans-Pacific record setter Clyde E. Pangborn for flying, and bush pilot Noel Wien for operations. It was a signal group that set a high standard. These recipients, called Pathfinders for their leadership qualities, would, in turn, enrich the museum stories of individual accomplishment and serve as motivation for students. The idea worked, from the first gathering in a hotel ballroom to the annual gala now held at the museum. It also augured well as a special event for the community and enriched the case for the museum and its appeal to donors.

In November 1982, First Interstate Bank provided a $500,000 line of credit to allow the project to continue construction of Phase 1. In effect, this was playing a development game of chicken with the schedule, budget, and debt. It was a strategy to move ahead without total funding in order to overcome the impression of a stalled project. There was little choice. It was move or remain stuck in place. The incremental steps showed action. The dodge worked as long as enough cash was acquired to keep the contractor working on-site. It was a struggle that became easier with progress. The Red Barn began to evidence its beauty.

Phase 1 rumbled along with no noticeable delays to the casual observer, and the Red Barn began to emerge as a magnificent structure of immense historical value. The environment changed for the museum with growing public enthusiasm and anticipation of its opening.

This first phase was memorable in not only saving the Red Barn but doing it in a way that gave it a new purpose as an interpretation of flight history and technology. Phase 1 was

Phase 1 began with restoring the Red Barn over a new foundation.

conceptualized to feature the Red Barn with an administrative and public-service wing, a modest but cost-effective operation and a good bit of product for the money. This was something called a museum, and even though just parts of what was to come, it had to prove the concept and leverage the rest of the project. This was a big chore.

T. A. Wilson

Thornton A. Wilson rose to lead Boeing as it teetered on the brink of disaster. The drastic action he took at the start of the 1970s saved the company, which he led back to brimming health and record profits.

Dramatic as this statement sounds, it scarcely evokes the fascinating man known simply as "T" to his colleagues and acquaintances. Wilson was born in 1921, a Missouri farm boy raised with self-reliant Midwestern values. Receiving a degree in aeronautical engineering from Iowa State University in 1943, he moved that year to Seattle—his fiancée Grace's hometown—where he went to work for Boeing.

World War II was then in full swing. Boeing's B-17 and B-29 programs needed all the engineers they could get, but Wilson soon found himself assigned to the elite team creating the revolutionary B-47 Stratojet. Later still, he would serve as project engineer on the B-52 Stratofortress program.

T. Wilson at Boeing, circa 1950s.

Amid a sea of engineering talent, it was clear that the young Missourian was exceptionally gifted and capable. Being encouraged to continue his education, he departed in 1946 and returned two years later with a master's degree in aero engineering from the California Institute of Technology. He was named a prestigious Sloan Fellow in 1952— the fast track for senior management at Boeing—and spent a year learning industrial management at the Massachusetts Institute of Technology.

In 1958, Wilson led the Minuteman program to develop an intercontinental ballistic missile system for the U.S. Air Force. Crucial to national security during the Cold War, Minuteman was an enormously complex and challenging program. Under Wilson's able leadership, it succeeded brilliantly, further enhancing Boeing's reputation as a world leader in large-scale systems integration.

Because of these successes, T. Wilson was appointed company president in 1968, CEO in 1969, and chairman in 1972. There were then four commercial airplane programs under way: the 737, 747, U.S. SST (later canceled), and a redo of the 727. The company might have gotten away with being so financially overextended had a recession not hit, causing airline orders and associated revenues to dry up.

Suddenly Boeing was gushing red ink. T. Wilson now faced the most trying time of his entire career. Massive layoffs alone could save the company, but their economic consequences would devastate the Boeing community and the greater region. Few leaders would have had the courage to cut deeply enough; those who did might not have known how to prune the tree without killing its roots.

Wilson laid off some 60,000 employees, leaving just 39,000. Thanks to his quick action, the company survived. He himself paid a price, though, experiencing a heart attack at age 48. Recovering, he rebuilt Boeing and launched the 757 and 767 in the late 1970s. Together with the 727, 737, and 747, these advanced fuel-efficient twinjets helped bring record profits by the middle of the next decade.

Tall, imposing, and forthright to the point of bluntness, Wilson always gave clear direction. As Boeing's chairman, he cared little for the trappings of power, dressed casually, and drove an economy car. Not sentimental, he did not immediately see the value of relocating and restoring the Red Barn, but came around when he realized the degree to which the community cared and how much it had accomplished. He played a central role in raising funds for the Museum of Flight's Great Gallery, which today bears his name.

T. Wilson stepped down as CEO in 1986 and retired as chairman the following year. He died in 1999 at his Palm Springs home at age 78.

Top: As early as 1983, the museum held air shows at the new site.

Bottom: Project architects Lidija and Ivo Gregov discuss the Red Barn restoration with Bill Allen.

By February 1983, McCann Construction had substantially completed the Red Barn refurbishment and administrative building, and the following month, staff relocated from the First Interstate building to the new museum to manage the installation of exhibits, set up a capital funding support division, and prepare to meet, greet, and educate. Although the facilities were modest in size, the restoration of the Red Barn was an incredible testimony to what was to come. Its image was magnificent and spoke to what this museum was all about, with additional site improvements creating a lot of anticipation.

The restoration design and engineering were thorough under the direction of Ibsen Nelsen's project architects Ivo and Lidija Gregov, harking back to a key 1920s version of the building, but making it safe and comfortable for its reuse as a place of public assembly. What appears to be simple construction contains a lot of details in its bones.

The foundation on the compacted fill material is pile-supported reinforced concrete. Exterior wall restoration incorporates insulation hidden with new exterior siding and finish. The Red Barn incorporated new exterior windows with insulating glass, roof insulation, and new metal roofing, as well as structural bracing and a fire-suppression system. This was done to meet contemporary energy standards, and earthquake and structural codes. A lot of attention was given to maintaining the simple, elegant look of the structure, while bringing it up to code. This did not have to be done, but it was deemed the appropriate way to restore. It was more difficult and more expensive, but the right way to go.

Criticism was heard from the historic-preservation community for some of these features, but the museum team had determined that these provisions were necessary and proper, even sacrificing a small preservation grant in the process. This approach was validated when, in 1984, the project won a regional historic preservation award.

The attached portion of the new administration building was joined by a two-story glass-enclosed connector featuring a public entry. This addition allowed for a store, offices, event room, and public amenities. It also housed a robust HVAC system for a controlled environment. As Ibsen Nelsen's architects explained: "The new structure is intentionally industrial. Construction and finishes are simple and durable, with attention paid to minimum maintenance and long life."

In July 1982, Promotion Products Incorporated of Portland, Oregon, was contracted to prepare final design and fabrication for the Red Barn exhibits. The gift shop design was included within the exhibit contract to make this shopping space an interpretive experience, similar to what Lovering had seen at the Nut Tree in California. With the museum board adopting the exhibit story line and issuing the contract, all design segments for Phase 1 were in progress.

The Red Barn exhibits were conceptualized to introduce flight themes within the restored historical space. They introduced visitors to the principles of flight and the history of aviation. A section titled Vision traced the path from the Wright brothers to the European leadership era, and back to growth in this country, with the attendant role of The Boeing Company. Associated themes covered the golden age and building for war, with its influence on evolving commercial transport. The industry was traced from early days, focusing on crafts and processes practiced at Plant 1. These exhibits allowed the beautiful Red Barn interior to come to life with sight and sound and color.

The layout included what was called the Pacific Aero Club, a small theater and meeting space honoring the seminal 1915 Seattle club of enthusiasts. A gathering space accommodated functions, which helped move the institution into the hospitality business.

Rocketry was interpreted to set the stage for more space themes to come in the Great Gallery, and there were special displays of airplane design, flight training, technology, and production crafts. The former chief engineer's office on the second floor was restored with mementos of former Boeing president and chair Claire Egtvedt.

Top: The exterior of the museum as it opens its doors after the completion of Phase 1.

Bottom: Red Barn interior, circa 1983.

Paul and Lucy "Lu" Whittier with their donation as the "Naked Jenny" is prepared for display.

One of the most beautiful exhibits in the Red Barn was the Curtiss JN-4D "Jenny." Paul Whittier and his donation of the valuable aircraft make for a landmark story in the formation of the museum. Museum staff member Frank Koral introduced Whittier to the museum in the early 1980s. Whittier, who had flown Jennies as a young man, loved aviation almost as much as sailing or having a gimlet with a friend. He was an unlikely donor target, as he had grown up in California. He was an heir to the Belridge oil fortune and the son of one of the founders of Beverly Hills. His philanthropy was mostly in California, where he had his principal residence, and his only local tie was another residence in Friday Harbor, Washington.

Whittier wanted to be part of the museum family of supporters by doing what might not otherwise be done. He funded and assembled a top-notch team for the "Lucy T. Whittier Storm Door and Flying Machine Company," and that talented bunch produced the magnificent Jenny, one of the finest restorations on display in any museum. What was so important about this is that it fit perfectly with the museum strategy of development phasing, which was to build a fine institution, and in turn, attract a collection. The excellence of the Jenny answered any questions about the commitment to quality and craftsmanship. One of just two aircraft on display in the Red Barn, it was a star in what was a small, off-Broadway production, and it is not an exaggeration to say it helped to put the museum on the map.

As construction progressed, excitement began to build in the community. This was a time when the locals realized that what they had heard about for many years was becoming real. This general enthusiasm was further leveraged in the local media. Among the most prominent was a series of mini-documentaries entitled "Moments in Flight," produced by KOMO-TV to herald the forthcoming museum opening. These 12 short segments, in 30-second and one-minute lengths, "capture the mystery, excitement, and romance of human flight," according to the station. Local producer-director Michael DeCourcey, who had developed a similar series of

"Tut Minutes" for that Seattle Art Museum blockbuster, was retained for the series. Critics praised these vignettes. One wrote that they were "technical, sometimes they are historical, other times they're just plain fun. But at all times, they do invoke the magic of flight and man's fascination with it." This was a high-budget production of top quality in writing, cinematography, and acting. The mini-stories, about airmail pilots, the Red Barn,

the B-17, or space exploration, grabbed the attention of viewers and created anticipation for the museum opening. DeCourcey and crew won a slew of Emmys and video awards for these artistic productions, which hold up well more than 30 years later.

Phase 1 officially opened on September 1, 1983, with a modest and short ribbon-cutting. Governor John Spellman arrived at the ceremony quite shaken, as all were just learning of the death of U.S. Senator Henry "Scoop" Jackson.

Although spirits were dampened, the governor spoke with pride on this momentous occasion, as he had participated in the project's circuitous route from idea to reality while serving as King County Executive. Bill Boeing Jr., with the assistance of Emma Brackin, widow of the beloved founder, cut the ribbon and the doors were swung open.

Top: Bill Boeing Jr. speaks at the Red Barn ribbon cutting, September 1, 1983.

Bottom: Emma Brackin and Bill Boeing Jr. cut the ribbon with Governor John Spellman (right) and King County Executive Randy Revelle (left).

This was show time. Could this museum—with limited exhibits, few aircraft in and out, and a modest budget—attract visitors and serve the public? Could the museum operate in a manner that would justify the capital drive to finish the complex?

This was a dicey period during which, in Captain Jack's phrasing, it was important "not to stub your toe." While the museum operated to budget and generated earned revenue, it also managed imaginative programming and posted unexpectedly good numbers. The early operational phase could have fallen short and stopped support of the project in its tracks. Almost as bad, there was also the possibility that Phase 1 would be just good enough to get a community attaboy and no additional support. Neither of these occurred.

That it worked so well is a testament to the remarkable restoration of the Red Barn, special events, a fast-growing support constituency, and, ultimately, the efforts of the board, staff, and volunteers.

Visitors on Opening Day of the Red Barn, September 1, 1983.

The learning curve was steep and daunting. Anyone in the museum business can warn of the challenges of daily operations, and few are prepared for the complexity of merging these functions with continued planning and fundraising.

NASM consultants had warned about the sudden onslaught of opening the doors for something so eagerly awaited, a transformation that often results in disorientation. This is a common phenomenon, as the board and staff are tired, and some take time off or move on to another position. After the initial euphoria, daily functions can move the needle closer to drudgery. Moving into operations requires a different kind of mentality and is as challenging as building but requires shifts in skills and personalities.

Fortunately, the museum did not suffer from these common maladies, probably because the job was not done. It transitioned into operating, attracting visitors, producing popular programs, educating, gaining membership, and training volunteers while still planning and funding and building. Although the staff was learning by doing, they were also smart, eager, energetic, and imaginative.

Staff, board, and volunteers had to prove they knew the right way to run a successful museum, and they were collegial and creative in their approach. Meetings were not about administrative rules or arcane museum topics, but about how to attract, entertain, and satisfy the visitor. Suggestions were solicited from all, and the team came up with a variety of program ideas—some brilliant, many questionable, and some never to be mentioned again.

The operating structure was thin in levels of management. Employees were generalists, capable of stepping into other positions as needed. Communications were continual. Whether it was a scheme for increasing the member base, acquiring an artifact, or putting on the greatest show on Earth, everyone with a good idea had an easy path to the director's desk. Mistakes were made and lessons learned. Each day was a new experience, producing ideas for growth and expansion. The board provided support and let the staff do its stuff, while volunteers pitched in with enthusiasm, forging a team effort.

From the start, an invaluable skilled cadre of docents and other volunteers buttressed the museum. They took to the tasks of opening the doors, welcoming an unexpectedly large number of visitors, putting on public programs, and staging memorable special events. This was team spirit that promoted the sharing of skills and ideas.

The new model demonstrated proof of concept in spades. The active airfield environment was a dramatic backdrop just as expected, and it offered up daily flight dynamics. The enjoyable programs and events rapidly gained attention. The high quality of historic restoration and museum design signaled that this was a classy project, and that boded well for future development. This quality also inspired other airfield improvements that slowly formed a cultural environment in the industrial area.

Opening Phase 1 attracted artifacts, helped to build an archive, recruited partners, and established the museum as valuable to the community.

This success propelled the capital campaign and brought in Boeing corporate and employee support that reached aerospace suppliers nationally, attracting the generosity of regional foundations and major individual donors. The pace of new funding accelerated. Using the phasing strategy, the Museum of Flight came out of the blocks quickly and began a fast pace for development.

Proud of their new museum: Trustees Bill Boeing Jr., Dick Bangert, and Wells McCurdy with Senator Henry "Scoop" Jackson (second from right), shortly before the museum's opening.

The Red Barn shortly after the opening of the Museum of Flight.

CHAPTER NINE

Corrigan, Ford, and Concorde

The early operations of Phase 1 were risky, and no one could easily predict what would transpire. Absent an endowment or operating fund, the challenge was to run on net revenue gained from admission fees, annual memberships, gift sales, and additional earned income. The fact that only a few museums in the country achieve this performance level made the test even more daunting.

While performing this economic balancing act, the organization was tasked with offering colorful and popular programs and events that could keep the project front and center. The board, staff, and volunteers hunkered down and proved the concept in a decisive manner. Making this directive even more difficult was a commitment to keep fees modest, starting at $3 for an adult.

In the first full year of operation, the museum attracted 127,000 visitors, a number that exceeded that of larger museums of its type around the country as well as more established museums in the region. More than 15,000 students participated in the informal education programs, delivered both in the museum and through outreach.

Membership, not counting the inflated number from the Boeing employee capital campaign, grew to more than 6,000, with a high proportion of those early supporters showing interest in participating and volunteering. A robust cadre of 200 volunteers began training to augment the full-time staff of 15.

Earned income comprised 30% from gift sales, 26% from membership fees, and 24% from "other." The 20% revenue from admissions was lower than the other categories because of the low price of a ticket. The unlikely mantra was to keep a visit reasonable in cost, make it fun and educational, and pay the freight from the gate.

The organization reviewed and adopted a set of operating and governing policies that codified the guidelines with best museum practices and updated the charter and bylaws to meet current requirements. The governing body was enlarged to provide for new trustees in support of the expanded operations and a newly launched capital campaign to face up to the impending surge to fund the Great Gallery. The Executive Committee of the Board instituted operating policies and new business systems to guide management and track sustainability.

Santa (Bob "Swage" Richardson) arrives in his Grumman Goose for the Forgotten Children's Program, 1983.

The basement of the Red Barn provided a large classroom, a multi-use space, and staff offices. A break room gave adequate space for volunteers and some of their gear.

This education center was organized so that several classes could visit at once, staggering the use of the classroom with tours of the exhibits, films in the small theater, and activities on the grounds. The facilities worked well for receiving the students, getting them comfortable, providing docent-led tours of the exhibits, and organizing aerospace-themed activities. The multi-purpose room was a place to leave coats and even to sit down to a sack lunch. Of course, all of this was enhanced by the museum's location on an active airfield with something going on all the time.

Various groups would volunteer to demonstrate the art of model making, throwing a boomerang, flying kites, making paper airplanes, any activity that allowed a student to join in and be exposed to principles of flight. Programs were not sophisticated but they attracted volunteers, and each was enriched with history and science. Best of all, these programs grew in popularity with students and their teachers.

Special events attracted thousands and began to build the brand, further establishing the museum as a place of excitement. Hundreds of disadvantaged children were thrilled at the December arrival of Santa (Robert "Swage" Richardson) in his Christmas Grumman Goose.

Robert E. Richardson

Bob "Swage" Richardson was important to the early building of the museum in several ways. Admittedly not a museum person, Richardson believed that aircraft should be flown and not hung up or stored. That was his take, and it is consistent with that of a lot of others who have done so much to save great aircraft, restore them, and fly them for public enjoyment.

Swage in the left seat of his Boeing B-17F.

Flying was his passion, or at least the foremost of a number of passions, including jazz music, sailing, and socializing. He played the standup bass and loved jazz clubs. Over his career, he had developed several marinas and was adept at boating. His hands-on experience and business interests met when he took over University Boat Mart from his father, evolving several interests into University Swaging. The company developed into a very successful "swaging" business that gave him his popular nickname and kept him involved as a subcontractor to the aerospace industry. His company earned a United States Small Business Award in 1977 and 1982 as well as several awards of excellence from The Boeing Company. This success allowed Richardson to enlarge his air force, which included a number of planes, finally resulting in a Howard DGA and a Grumman Goose.

He flew these not as some kind of isolated treasures, but as daily drivers that could take him in style where he wanted to go. He often told of loading up the Goose with a few cases of Scotch, his fishing gear, and a few buddies and taking off for the lakes of British Columbia. Richardson was popular, enigmatic, fun, complicated, and at the same time, very down to earth. He was one of a kind.

It didn't take much coaxing to get him to purchase a B-17 in 1984; it was the kind of thing he would want to do for the museum, and it was a high priority for the institution. Locating the B-17F for the museum was a dream come true for Swage, who searched Europe and North America to find this special Boeing-built flyable F model. He worked hard to restore the historic aircraft, flying around the country to display it and to promote the museum at air shows. He made a lot of friends for the airplane and for the museum. In 1989, he flew to England, where the museum B-17F played a starring role in the movie *Memphis Belle*, which captured the role of this great bomber in WWII.

Lovering, a longtime friend, described Swage as "the best kind of friend and aviation enthusiast a museum could ever hope to have. We hope he has had half as much fun finding the B-17 and bringing it home to Seattle as we've enjoyed through our association with him."

Swage passed away unexpectedly and much too young in 1990 at the age of 59, survived by his mother and two sisters. His memorial service at the Museum of Flight on April 8 of that year witnessed a throng of friends filling the galleries. The announcement of his generous arrangement to make sure the B-17F was transferred into the collection at the museum came as a surprise to no one. This is an airplane of a thousand stories, and of special meaning to the museum in its first 50 years. Bob Richardson is in the same category.

More than 30,000 visitors attended the three-day Annual Salute to Aviation, which brought in military, commercial, and vintage aircraft to the airfield environment for interpretation. Occasions of this kind proved the museum had a viable model in its event-driven approach to operations. The museum took the best qualities of air shows and reduced them to a manageable scale. These events were successful in building an audience and gaining local and national media attention.

The museum was always in search of some fascinating event, dramatic enough to attract visitor attention and capable of being staged on the cheap. In 1984, someone tossed out the possibility of an event to celebrate the 46th anniversary of Douglas Corrigan's 1938 solo transatlantic flight. He had flown in a Curtiss Robin, and there was one on exhibit in the Red Barn. That was enough of a connection to merit taking a serious look.

Corrigan's flight was the culmination of a flying career as a barnstormer and airplane builder. He had worked on the *Spirit of St. Louis* and admired Lindbergh and his achievements. For several years he had tried to get permission to fly his Curtiss Robin transatlantic, but his requests were denied. He was lucky to keep the old beat-up aircraft licensed, and the FAA had grounded it several times.

Douglas "Wrong Way" Corrigan with his Curtiss Robin, 1938.

Corrigan was logistically unprepared for a transatlantic flight, though as a master mechanic, he kept the power plant in his well-used Curtiss Robin in excellent condition. Everyone he talked to agreed on one thing: Corrigan's airplane could not carry enough fuel to cross the Atlantic. He had no special-built equipment, no support team, and no financial backing. What he did have was a $10 leather jacket, a $2 Woolworth shirt, a $3 pair of Sears trousers, sturdy new $3 shoes, and a necktie someone gave him. He crammed the $3 he had left into his new trousers.

On July 17, 1938, with a flight plan to return to Long Beach, California, he took off from Floyd Bennett Field, New York City's first municipal airport, and just kept flying east. He had not fixed a leak in the gas tank and developed a puddle of fuel in the cockpit while over the Atlantic. With no radio, an out-of-date compass, and few provisions, he somehow made it to Baldonnel Aerodrome in County Dublin, Ireland, on July 18 after a 28-hour flight. His provisions had been two chocolate bars, some fig bars, and 25 gallons of water. Corrigan shared his leftover staples with his surprised hosts in Ireland.

Corrigan constructed a story that he had missed some markers and somehow gone the wrong way. Anyone who knew him and his flying ability knew this was not true, but Corrigan stuck to the story until his death in 1995.

After his transatlantic flight, he was honored with a ticker-tape parade in Manhattan, which more people may have attended than the epic Lindbergh parade, explained by his Irish heritage.

He retired in Santa Ana, California, and his celebrity faded. Most who called him were trying to locate his aircraft. Museum staff were warned that he was reclusive and did not trust many outsiders, but a call was made with an offer to pay the expenses for his appearance at a special event. Corrigan countered with an offer to visit if he could sell his book and keep the proceeds. He agreed to journey up in July.

The museum produced the Wrong Way Corrigan event out of its petty cash fund. Due to his contempt for young pilots and smokers, he arrived by train on July 13, 1984. Picking up on his arrival by rail, local media took a particular interest in this event, and it was covered in a way it might not have been otherwise.

A room was reserved for Corrigan at the airport Hilton, which put his nameplate on the door, although he asked for the hotel expense in cash if Lovering would allow him to crash on the couch of the director's small rented houseboat. Corrigan seemed disappointed not to make that deal.

The museum requisitioned an antique auto that was the same vintage as the one that had carried him in the Manhattan parade. Corrigan rode the wrong way on the one-way Fifth Avenue in downtown Seattle, sitting on the back of a 1937 Cord convertible, wearing his original flight jacket with several cracks on the back sealed with clear packing tape, and thoroughly enjoying a recaptured celebrity.

Top: Wrong Way Corrigan arrives in 1984, posing here with the Museum of Flight's Curtiss Robin.

Bottom: Corrigan heading the wrong way on Fifth Avenue to celebrate his 1984 visit.

Corrigan was a nonstop attraction at the museum for the next few days. He did not care for the lunches and dinners provided, as he lived on candy bars and orange soda. But his energy was formidable as he continued his "wrong way" fabrication in the stories he told to the crowds who came to the museum to meet him, buy his book, and get an autograph.

He was a lot of fun, though considered by some a strange dude. This event in Seattle rebooted his legend, and he had several more years of stunts here in the states and abroad. Corrigan was a showman, attracting attention and knowing how to deal with it. The excellent local media coverage was good for his image and for the museum.

Lovering said, "Those of us who spent time with Douglas Corrigan during his visit developed abiding respect for him. He was a small, sprightly, outspoken, boyish gentleman, wearing all of us out with his enthusiasm and energy. It is easy to believe his flying feats; one senses he could still pull off a stunt or two."

148

On the day that Corrigan was to return to his home in Orange County, Lovering drove him to the train station. At his request, they first visited the Space Needle. On the observation deck, Corrigan marveled at the landforms, buildings, and arterial system, exclaiming, "You know, if they'd had structures like this in the city, I would never have made it as a barnstormer." A lot of his early flights were made to take a look at farms and towns from the air.

At the station, Lovering was pleased to find a dispenser with orange soda, an excellent way to start a long trip on a hot day, at least for a 77-year-old vegetarian who looked at regular meals as dangerous contributors to the aging process. As Lovering left him, Corrigan was wearing his old leather jacket from stunt-flying days, sipping soda pop, in animated conversation with the young woman in line ahead of him.

Lovering remembered this event and this new friend fondly. "Over four days I heard Douglas Corrigan tell his 'wrong way' flight story over and over with consistency in detail. He didn't tell much about his long barnstorming career, helping to build Lindbergh's *Spirit of St. Louis*, the war years, or his preeminence as a test pilot. He relished the role of court jester, the title 'wrong way,' the humor of his navigational error. Few, if any, believed Douglas Corrigan was trying to fly to Long Beach on July 17, 1938."

During this event, another stellar production was in the works. Trustee Jim Curtis, an avid golfer, had promoted the idea of a celebrity tournament to focus attention on the museum. He

Former President Gerald Ford at the Museum of Flight Pro-Am golf tournament, 1984.

enlisted friend and fellow golfer Clem Powell, a Seattle businessman, and others to organize a prestigious Pro-Am tournament for the course at Sahalee Country Club in Sammamish, Washington. Club pros reached out to national professional golfers, while the organization got a favorable response from former President Gerald Ford to participate. With a great response from participants and sponsors, the tourney caught on in the community.

Arriving at Boeing Field in a Weyerhaeuser corporate aircraft, President Ford was treated to a personal tour of the Red Barn by T. Wilson of The Boeing Company. On the evening of August 26, 1984, in the grand ballroom of the Westin Hotel, the crowd gathered for bidding on teams in which amateurs would play with a designated professional. The winning bid of $8,000 by four Boeing executives won the opportunity to play with Gene Littler and President Ford.

That evening's party was attended by dignitaries, including Senators Gorton and Evans and Governor Spellman. The governor remarked that "Kitty Hawk was where the dream was born, but it was fulfilled at Boeing right here." The event brought together a large slice of the community through the mutual passion for golf.

At the tourney the next day, the President was a blast, a good athlete and competitor, and by all accounts a great guy to be around. When asked about his game, he said, "It's getting better and the best evidence is that I am hitting fewer spectators." The event ran almost constantly in the media and garnered additional attention to the mission of the museum.

A supersonic event later in 1984 was initiated not by MOF, but by an enterprising local restaurateur, Mick McHugh. McHugh, then and today, is owner and proprietor of F. X. McRory's in Pioneer Square, modestly advertised as America's Number One Bar. McHugh is an ebullient and fanciful character who is often referred to as Seattle's official greeter. A one-time employee of legendary local restaurateur and funny man about town Ivar Haglund, McHugh learned well the lessons of community involvement as business promotion.

He and Tim Firnstahl, his partner at that time, ran several eateries and bars and never missed a chance for promotion. Their fun and games were a big part of the attraction of their restaurants, which also delivered good food and beverages and a wholesome feeling of family.

Top: Left to right: Malcolm Stamper, Gerald Ford, T. A. Wilson, Phil Condit, and Dean Thornton.

Bottom: T. A. Wilson (right) and Jim Curtis accept a capital campaign donation as part of the Pro-Am tournament.

Though McHugh's operations were much more about hard liquor and beer, he seized upon a wine stunt that perhaps was his boldest stroke of all. Reading the hype about the race to deliver the first release of Beaujolais Nouveau (BN) in France every November, he pondered the opportunity.

It was not a great wine. Its arrival was more of a seasonal event, and at any rate, quality of wine meant little to McHugh. As he said: "It is not a serious French wine. You can read the newspaper through it, if you will." He did understand, however, that its rush to market could result in a tremendous promotion in Seattle.

Scouting France in 1983, McHugh was particularly struck by the sight of the magnificent Concorde at the airport. The stunt came together immediately in McHugh's mind, and he decided he would charter Concorde and race with BN to Seattle. He mused that seeing Concorde up close just might be good for Seattle after its disappointing experience with the Boeing version of a supersonic aircraft. Boeing had been working on its 2707 model, an advanced concept of a supersonic transport, when Congress canceled its funding in 1971, contributing to the "Boeing Bust."

The Concorde lands at the Museum of Flight, November 16, 1984.

McHugh took one of his many chances and chartered the Concorde to bring in the BN at as fast a speed as could be recorded at that time. This event required an investment of $185,000, a lot more than most would spend on a marketing promotion. But McHugh was confident this was something good for his business and the community, and he plunged ahead. He understood dollars better than most and began to enlist passengers to help defray the expense. Falling short, McHugh trolled the community for partners.

He drew up a deal with Lovering to take on some of the expense and, in return, to have the aircraft at the museum and available for some short loop flights. The supersonic flights over the Pacific from and back to Boeing Field were named "Flights to Nowhere." Even with this inelegant name, they caught on with hundreds of people who otherwise might never have the chance to ride on Concorde. Both Lovering and McHugh began to see the financial light at the end of the tunnel.

An estimated 10,000 people waited at Boeing Field for the arrival of the exotic plane. McHugh was riding in the cockpit as the Concorde coasted from Mount Rainier toward the city. He remembered seeing Boeing workers watching from the roofs of the company's buildings and truckers watching from the tops of their vehicles parked along Interstate 5.

At least 5,000 more spectators were viewing the final approach and landing from nearby arterials, which stopped traffic. Observers, who never expected such a scene, left their cars in a massive jam and watched as this incredible aircraft landed at Boeing Field. The event was dramatic beyond even the considerable imaginations of McHugh and Lovering. It commanded attention from thousands of spectators who will never forget the occasion, from all of the local media, and from all of the major national networks.

Among the passengers who deplaned and spilled into the cheering crowd was Robert "Swage" Richardson. Swage had planned to be in Paris to attend a gig by a woman friend and jazz singer, and he could not think of a better way to travel back. It was a spectacular event. It did a lot for McHugh and his eateries, but it did a lot more for the Museum of Flight.

For several days, the Museum of Flight was official supersonic headquarters. Young and old alike streamed through the doors, learning more about this unique transport. On Friday, November 16, a free walk-around tour was provided to all students arriving on a school bus. Several thousand students from 60 schools between Bellingham and Olympia joined in the fun of getting up close to Concorde.

British Airways personnel joined museum staff and docents in presentations on the history and technology and service culture of the Concorde. Youths were kept busy with workshops on the dynamics of flight and the opportunity to make model supersonic aircraft.

Later that afternoon, the Flight to Nowhere took off with 100 passengers, including three students who had won seats in an essay contest promoted by the museum. Trustee Bruce McCaw took the fast, short hop along with his mother, Marion. The testimonials from those on that flight confirmed the significance of this event. Reporter Polly Lane reported that "in 2 hours and 39 minutes, including an hour and a half at more than the speed of sound, we streaked more than 1,200 miles west and south of Seattle, nearly halfway to Hawaii."

Boeing Company executives and employees joined the festivities. If there was any residual animosity over the loss of what most experts believe was a much superior Boeing SST program, such was not witnessed in the enthusiasm of the occasion. Boeing has always been about making airplanes, and this was a great one to see up close.

McHugh proved his promotional talents with this over-the-top flight, and it seemed to matter little to anyone that a local wine importer managed to get the BN in before the Concorde. That fact was lost in the smoke of the landing.

The museum proved again that the real operations of aircraft were among the best attractions, and this particular event was beyond their wildest expectations. With the extraordinary success of this promotion, the museum knew it would have to do this again and even expand on the occasion. Lovering and McHugh followed up with a formal written request to British Airways for a donation of a Concorde upon the end of its flying career. Staff also began to imagine how to bring back this Concorde for a repeat performance.

With Phase 1 complete, the Great Gallery was about to go from dream to reality.

CHAPTER TEN

Flying Paper, Flying Fortress

The new museum kept up steam, searching out events that incorporated programs, captured attendance, and attracted continuing media attention. The Second Great International Paper Airplane Contest was low-budget and high-profile, with global participation and coverage. It came along slowly and then catapulted into media awareness.

The special event department combed the aviation journals for subjects of interest and discovered *Scientific American*'s First Great International Paper Airplane Contest. This magazine-sponsored challenge resulted in articles and a book that promoted the hobby. The staff was aware of the popularity of paper airplanes from gift shop sales and various youth programming activities. The idea of a contest was pitched as something the museum could manage, and it seemed to work within the parameters of small budget and broad interest.

The idea was to re-enact the First Great International Paper Airplane Contest held some 18 years earlier. Staff contacted *Scientific American*, and they had no interest. But the idea was gaining ground and seemed too good to just let go.

The idea was next pitched to *Science 85* magazine as a partner and sponsor, and they were receptive. AG Industries, the company that made the popular WhiteWings airplane kits, was a natural and signed on as a funding sponsor. After a few inquiries, the National Air and Space Museum agreed to be a promoting partner, which gave the project credibility.

Science 85's interest was fanned by the fact that the proposed contest was easy to handle, with potential large scale. It also played to the fascination with aviation, with an opportunity to focus on the emerging technology of molded fiber construction. Paper is a composite material that can translate this new technology into a simple design, fold, and fly game. For the Smithsonian, the interest was in a hobby with a flight motif that could make for a fun and colorful exhibit. With enthusiastic handshakes the event was launched.

By May 1985, the museum basement was crammed with 5,000 entries from 29 countries. The packages were specially made boxes to house each entry, carefully wrapped, and annotated with instructions for flying. Some of the instructions were in elaborate detail, supported by beautifully drafted graphics. From the elegance of the packaging, it was clear that the contest had fierce competitors.

The Paper Airplane Book, published in 1985, is the official book of the Second Great International Paper Airplane Contest.

The contest rules were simple. Entries were to be made from paper, with glue and cellophane tape allowed for bonding. The four divisions of competition were distance, time aloft, aerobatics, and an opportunity for displaying aesthetics. Even those in the latter category were expected to fly at least a minimal distance. Entrant levels were professional (aerospace educators and engineers), nonprofessional, and a junior class for those under the age of 14.

The pilots, or more accurately the engines, were volunteers from the local chapter of the American Institute of Aeronautics and Astronautics who included aerospace engineers, educators, pilots, and a variety of flight professionals. It was entertaining, but serious.

The flying was all done under the supervision of a cadre of judges recruited nationally. The celebrity monitors were Ilan Kroo, assistant professor of aeronautics and astronautics at Stanford University, who also worked as a consultant to NASA's Ames Research Center; Michael Collins, test pilot and Apollo astronaut, who also served as a director of the Smithsonian National Air and Space Museum; Dennis Flanagan, editor emeritus of *Scientific American* magazine; Dr. Sheila Widnall, professor of aeronautics and astronautics at MIT, a local woman who later became Secretary of the Air Force and an MOF Pathfinder; and Dr. Yasuaki Ninomiya, who was at the time the world's leading paper-airplane designer, a technical engineer enamored of paper airplanes since childhood, and who had won several grand prizes in the First Great International Paper Airplane Contest.

This prestigious group had as much fun as all the others, enforcing the rules and enjoying the contest. They also had to make a few unexpected judgments. When a Stanford University professor sent along a paper-laminated Aerobie design, similar to but more elegant than a Frisbee, it showed promise of being thrown out of the Kingdome. The judges ruled it not an aircraft, but it was a complicated point, and the flying saucer received an honorable mention.

Distance was defined as the measure in a straight line. Elapsed time only required the aircraft to be airborne from launch to finish. Dr. Widnall observed that if there were some good way to measure the path of those aircraft that looped and circled, there might have been a different distance winner.

At the end of the competition, Japanese citizens had taken eight of the ten first-place prizes and many more ribbons as well. Tatuo Yoshida, a paper-airplane designer from Yokahama, traveled to Seattle to watch his "children" perform. They did not disappoint, winning first in professional aerobatics and time aloft and taking four medals in all. Mr. Yoshida was a professional hobbyist with 50 years of experience, including a win in the First Great International Paper Airplane Contest in 1967. Americans did score a first, in the distance category, by Robert Meuser, a mechanical engineer from Oakland, and by Eltin Lucero of Pueblo, Colorado, in the junior class.

Young Lucero, new to the hobby, entered a classic dart design that traveled smoothly for 114 feet and 8 inches. His became a human interest story when it was discovered that the paper he used was a discarded appraisal report that he had picked from the trash while with his parents as they cleaned an office building. His win earned him a trip to the museum for the formal award ceremony.

On June 9, 1985, the winners were flown to Seattle for the award ceremony, held at halftime of the Kingdome's other World Indoor Paper Airplane event. Three thousand spectators cheered as each winner received a specially produced Bernoulli Medallion, culminating with a group launch of the winning aircraft. Young Lucero and his parents had the trip of a lifetime, which resulted in a photo essay published in *World* magazine.

The media found gold with this event, from the small aircraft to the mega-dome, the foreign invasion, and the celebrity judges. Their coverage promoted the museum across print and broadcast media. All of the components somehow aligned to create one of the soft news events of the year, with some of the winning aircraft appearing on the *Late Show with David Letterman*, as he unceremoniously launched them into his audience. *The Paper Airplane Book*, published in 1985, is The Official Book of the Second Great International Paper Airplane Contest, co-edited by MOF event manager Alison Fujino.

Lovering launches his record-breaking flight at the Kobe Bowl, 1985.

In August 1985, Japan Airlines and AG Industries invited Howard Lovering and Alison Fujino to attend the Japanese National Paper Airplane Contest, held at the beautiful Kobe Bowl. Lovering served as an honorary judge with Dr. Ninomiya and others.

In an opening ceremony, the judges were challenged to a time-aloft contest with professional sling-launched paper airplanes. Lovering had no experience other than a few trials with the beautiful craft but had read Dr. Ninomiya's instructions on preparing and trimming the aircraft for flight. Wonder of wonders, his aircraft launched steady and straight, caught some whimsical updraft, and took off to a height just barely visible to the human eye. The judges timed the plane sailing in broad circles until it was out of sight, ending the clocking at just over two minutes. There was a hush, and then a loud cheer for the announcement that Lovering had just recorded the longest time aloft for a foreigner in Japanese contest history. For the rest of the carefully managed competition, Judge Lovering wore his sash and medal, receiving many and frequent bows of respect.

On July 2, 1985, the museum squeezed in a long-awaited ceremony between all of these major events. It was a colorful groundbreaking for the new Great Gallery. The occasion kicked off when "Swage" Richardson and his Boeing B-17 taxied to the site with a payload of luminaries, including Senator Slade Gorton, Representatives Norm Dicks and John Miller, County Executive Randy Revelle, Seattle Mayor Charles Royer, T. Wilson of The Boeing Company, and Dick Bangert, chair of the museum foundation. Among national leaders who could not attend, there were written messages of congratulations from President Ronald Reagan, Vice President George Bush, Governor Booth Gardner, and U.S. Secretary of Transportation Elizabeth Dole. In a short presentation, Wilson announced that "we have raised $20 million and want to wrap up the campaign by the end of the year."

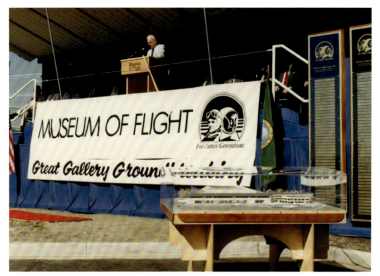

Top: Trustees and dignitaries get into the act at the Great Gallery groundbreaking, July 2, 1985.

Bottom: With a model of the Great Gallery on hand, T. A. Wilson opens the ceremony. The campaign motto: "For Future Generations."

Bangert shared a short review of progress over the years and general details of the new structure to be built on the site, with a promise to be open in 1987. He did explain that "we told the contractor not to disrupt the area too much right away because we're holding the 50th anniversary of the B-17 bomber and crews here." That largest of all events would follow within weeks, while the contractor was organizing his crew on the job site.

Two months after the paper airplane extravaganza, the Boeing Management Association sponsored this salute to the B-17 at the museum. Don Sachs, longtime BMA member and museum supporter, was a B-17 crew member and POW, going on to a management career at The Boeing Company after military service. Sachs, often referred to as "Mr. B-17," was known around the country for his commitment to military organizations and the history of the Flying Fortress. His affection for the Flying Fortress inspired him to do something special for the occasion of this 50th anniversary. In 1983, Don began recruiting a core group. He worked through the BMA to make this their priority, with a budget from The Boeing Company and partnership support from the museum.

Throughout three glorious days, from July 26 to 28 in 1985, more than 12,000 veterans and friends graced the museum property, joining 50,000 others to commemorate this great aircraft, its design and production, and its operational air crews who served America so well. Boeing chair T. Wilson officially opened the first-day ceremonies, held exclusively for B-17 crew members, and introduced General Curtis LeMay, who gave the keynote address. Also on hand at the opening were Ed Wells, who reminisced about designing the famous bomber at the ripe old age of 22, and Evelyn Egtvedt, whose late husband is considered the "father" of the B-17 for his early recognition of the need for such a bomber.

It was a nostalgic event, with flying warbirds, exhibits, special programs, and a convivial atmosphere that produced lasting memories with a reunion of spirit rarely experienced in our region. As General LeMay addressed the huge crowd, one of the visiting Forts flew overhead with a payload of five WWII Medal of Honor recipients.

Peter M. Bowers' book *50th Anniversary: Boeing B-17 Flying Fortress*, produced for this occasion, had sales around the world. Boeing Company artists created a series of B-17 paintings that graced the exhibits and were featured in the monograph. These paintings are some of the finest ever prepared for a museum publication.

The 50th anniversary of the B-17 commanded the largest attendance in the history of the institution. It was a fitting salute to one of the great warplanes in U.S. history and worked well on the airfield where so many had rolled out of production. This partnership with Boeing Management Association also returned the favor of their many years of support and proved that the museum property was a perfect venue to celebrate historic aircraft and their designers, builders, and crew members.

Top: B-17s were a regular sight at the museum grounds even before its opening. This B-17, *Sentimental Journey,* visited in the early 1980s.

Bottom: A large crowd at the Museum of Flight for the Boeing B-17 50th anniversary, 1985.

A valuable legacy of this tribute was the arrival that year of the B-17F that board trustee Bob "Swage" Richardson had purchased on behalf of the museum. This acquisition was planned earlier by Swage, and he used his keen sense of timing to bring it onto Boeing Field when there was intense excitement about that aircraft and its role in WWII. He found an F model, built at Plant 2 and rolled out in 1943. It was in rough shape after its military service and its years as a monument, crop duster, and fire suppression bomber, even a stint as a movie and TV star. It was flyable and just affordable at $275,000, so Swage wrote a check, passed his flight training, found an experienced pilot, and took to the air.

On a handshake with the museum, he agreed to restore the aircraft and jointly conjure up some flying events and a special "B-17 Bond" purchase to fund the maintenance and acquisition by the museum. These bonds were beautifully designed as keepsakes. They sold for $17, the B-17 book thrown in with a purchase of 10, and a grand prize of a one-hour flight on the B-17 with Swage for a purchase of 100. At the 50th event they sold well, with Swage flying several volume purchasers.

We Are Accredited

In 1985, with just a few years of operations under its belt, the museum applied for accreditation from the American Association of Museums (now American Alliance of Museums). This was a bold and unlikely application for several reasons. First, the institution, open for just two years, did not meet the minimum five-year operational requirement for accrediting. Second, the physical facilities and collection were limited at that time. If the gamble failed, it could be a setback to continued fundraising. Though it was not an orderly action, there were some compelling reasons for requesting this recognition.

The new museum had to address questions about its operating experience. Was the museum doing the right things? Confident in the operations, the request for accreditation functioned as a self-assessment. This assessment was calculated to find deficiencies to correct. If successful, the recognition would help to make the case to donors and supporters.

The museum submitted a preapplication asking that the five-year rule be waived in recognition of its past experience at the Seattle Center property, which was a stretch. Surprisingly, AAM concurred, and the laborious application process was started. The gamble was that the museum programming was more than adequate to meet national standards and that the ongoing construction of the Great Gallery would help evaluators visualize how the physical plant would work.

Staff went to work on the stacks of forms necessary to complete the formal application, a process that took nearly a year. In May 1986, the museum received word that it was accorded full accreditation. Along with the much older San Diego Aerospace Museum, the Museum of Flight was one of the first two flight museums to be so honored.

AAM accreditation brought national recognition and regional respect to any honored museum for its commitment to excellence, accountability, high professional standards, and continued institutional improvement. AAM's museum-accreditation program is the field's primary vehicle for quality assurance, self-regulation, and public accountability. It strengthens the museum profession by promoting practices that enable leaders to make informed decisions, allocate resources wisely, and remain financially and ethically accountable, all in order to provide the best possible service to the public. The early accreditation definitely enriched fundraising opportunities that needed the boost.

In May 2009, the Museum of Flight received reaccreditation by AAM, referring to it as the highest museum honor. Only 775 museums of the nation's estimated 17,500 are accredited. The Museum of Flight is one of 13 museums, and the only flight museum, with this distinction in Washington State.

The B-17 acquisition, though still a work in progress, remained an important story when the enthusiasm of the 50th anniversary waned. Many look to this event and this asset of a rare bomber as a turning point for the museum, the time when it became apparent that this could be a truly grand institution.

Immediately after the B-17 anniversary, on Monday morning, July 29, 1985, heavy equipment moved onto the property. With more than $20 million committed toward the goal of $26.4 million, the capital campaign was steaming ahead.

Behind the scenes, the museum's attorney, B. Gerald (Gerry) Johnson, was working feverishly on a new agreement for land tenure and project construction financing. Johnson had suggested this new form of public-private cooperation earlier in the year, and there was a rush to meet this deadline. The King County Museum of Flight Authority (KCMOFA) was a bridge to general-purpose government that offered a

way to shore up the landlord-tenant relationship with King County while also opening up better routes for construction funding. The Authority structure was in use for other community projects such as Pike Place Market and Pioneer Square, but this was the first for museum development. The hurry was to organize the Authority before the end of the year to avoid potential changes in the new federal tax-reform legislation that promised quite an effect on nonprofit tax-exempt borrowing. In addition, Johnson devised a scheme for a longer lease period, up to 99 years, allowed by the legislation that set up these authorities. The opportunity was too good to miss.

Top: The museum's prized B-17F remains one of the most important artifacts in the collection.

Bottom: The 80A was barged to Vancouver, B.C., to appear as a centerpiece in the Transportation Plaza at Expo 86, becoming the museum's first international exhibit.

Top: Trustee Bob Bateman (left) with T. Wilson in the Red Barn.

Bottom: Digging down before building up. Museum attendance continued to increase despite the ongoing construction.

The King County Museum of Flight Authority was formally established by action of the King County Council in December 1985. The new seven-member Authority board would consist of three individuals from MOF, three appointed by the King County Council, and an at-large member to be selected by the others. This went into effect with trustees Dick Bangert, Wells McCurdy, and Bob Bateman initially representing the museum. The Authority has always had representatives who cared first and foremost about the museum.

With the transfer of the museum and food-service parcels to the Authority, the door was opened for the museum to take on the restaurant project. The Authority helped to put together a $15 million financing scheme with Rainier Bank, a loan predicated on the capital pledges of $20 million. All of this work was compressed into a few months due to the deadline.

The museum released a media advisory on behalf of this county-museum partnership: "The legislation affords an improved situation for both King County and for our museum foundation. The museum realizes the tangible benefits of securing the most favorable construction financing while also improving its property lease. At the same time, King County is afforded improved property management by way of this new single-purpose authority dedicated to supporting the museum project."

With this bridge financing and the selection of the general contractor, the museum was in a position to authorize the completion of Phase 2. This was the start of the 18-month construction period. The building of the Great Gallery was under way and this was a heady time, when so many things were coming together.

Another moment of national exposure occurred by chance on October 25, 1985, when *Today* show weatherman Willard Scott visited Seattle. Scheduled for his spot from Pike Place Market, on a tip he hastily added the Red Barn. Cameras were set up at 3 a.m. so the show could broadcast at 4 to fit the 7 a.m. news slot in New York. There was a light rain, and the chill made for vapor trails when exhaling, but it was a cheery group that gathered around the inimitable weather guru and his constant chatter.

Knowing that Scott hailed from her neck of the woods in Virginia, Jane Brammer, a museum staff member, gifted Scott with a homemade peanut raisin pie. Somehow, 82-year-old museum volunteer and former Red Barn worker Carl Gustafson crawled out of bed to serve as a sidekick to Scott. It was hilarious as the soft-spoken Gustafson proved so charming and funny that Scott was reduced to tears. "Gustafson, that's Swedish isn't it?" Scott asked. "No, I'm from Missouri," Gustafson replied. Proudly wearing his red jacket covered with airplane memorabilia and insignia, Gustafson offered up some of the history of the old building as Scott held forth on the weather. It was just a few minutes of banter and some reporting, but another instance of the Red Barn as a useful backdrop.

Meanwhile, the museum worked day in and out to receive and educate and build an audience. In 1985, the second year of operation, the museum enjoyed 168,000 visitors, increasing numbers by 32% over the previous year. The education program served 14,000 students with a variety of learning experiences. Membership continued steady growth to 6,500, which also served up more volunteers, who contributed in many ways. The first traveling exhibit on the Wright brothers' photographs was well received. The strategy was to function as a full-service flight museum within the modest facility and limited display of artifacts. Highlighting the active airfield and offering dynamic programming made it possible.

Members of the campaign and operations staff in 1984.

Archivist Anne Rutledge (sitting) and volunteer Cande Dickson in the Red Barn basement archives, May 1984.

The young museum's artifact collection grew much faster than anticipated. By the second year, there were 44 major artifacts, aircraft and rockets, and 27 historic engines. The archives attracted donations of private collections. The 7,000 hardbound books, 125,000 magazines and journals, and 50,000 photographs in storage were content for the library and archives planned for the expansion. The strategy to construct the physical plant first and build it well, then attract the stars to the collection, was working.

Early operations were a successful test of the case for the larger capital campaign. Donors observed the initial results and realized the return on investment, and some early donors gave again or increased their pledge. The public response to the museum was beyond expectations.

The Phase 2 capital campaign gained momentum in all sectors, creating a more compelling case for T. Wilson and his team. The operating success attracted more support from the local business community at a time when T. was implementing an international solicitation of suppliers for the expansion. Each of these reinforced the other, and there was growing confidence that this large task had a good chance of success.

As Gus Grissom is credited with saying about funding the early space program: "No bucks, no Buck Rogers." The museum had proven its case for expansion. Now it was time to get all the bucks.

Let's Do the Numbers

The ambitious goal of being self-sustaining with net earned income was nearly met by the second year of operations. In the first year, 1984, earned income produced 87% of the budget of $462,000. This was accomplished with an adult admission fee of just $3. This was excellent performance during a capital campaign in which every dollar counted. It showed great promise for a strong fiscal future.

In 1985, with 91% of budget coming from operating revenue, the museum came close to its objective to run the place at the gate. Revenues increased 51%, with all sectors contributing. Membership income increased 31%, admissions 26%, the gift shop, showing great potential, some 99%, and the fledgling event-space rentals boosted the "other" category a whopping 200%. Net worth increased from $1.2 million in 1983 to $23.5 million with the success of the capital campaign in 1985.

With this showing, it was becoming clear that the museum had an operating model that was viable. This was good news to the campaign leadership, who so often and rightly heard this question from donors: "How are you going to sustain the place?"

The Museum of Flight gift shop, circa 1983.

Left: Clear-span
construction on the
Great Gallery,
May 16, 1986.

Right: Executive
Director Howard
Lovering (left) with
architect Ibsen Nelsen
at the Topping Off
celebration, 1986.

Above: A view of
Boeing's Plant 1 along
the banks of the
Duwamish River in
Seattle, 1968.

Right: Economic
hardship in the late
1960s was hard to
take lightly. It was a
challenging time to be
breaking

CHAPTER ELEVEN

Test Flight Great Gallery

Groundbreaking sent out the signal that the Museum of Flight was coming on stream with full force. At this point, the museum became a reality in the eyes of the community, the donors, and all who had worked so hard over the years. It was happening.

The Wilson capital campaign was successful enough to encourage the museum board to move ahead with construction financing. By late 1985 there were enough promises to join formal pledges and actual cash for the campaign team to see the end. This allowed for a funding scheme to get digging and compress the building schedule.

Gerry Johnson and his colleagues developed an imaginative financing package that attracted a consortium of local banks to provide adequate funds for construction to begin. These funds were made available at a favorable rate that saved a lot of money. The board concurred with this scheme and moved into bidding. The Gilbert H. Moen Company of Yakima, Washington, was selected and began almost immediately organizing the site and starting the excavation for the lower level. In just months, the contractor had 839 pilings in place, clusters of support for the Great Gallery structure. This large hole and the frenzy of workers said a lot for progress.

The groundbreaking ceremony had attracted a lot of media coverage, and the interest continued as the project advanced into construction. The unusual structure captured attention, and all phases of construction were monitored. As steel rose, so did the spirits of all involved. Clearly, this was going to be a monumental structure. The museum invited members and donors to a Topping Off celebration on August 6, 1986. The party, with refreshments, music, and tours, welcomed over 300 guests and drew substantial media coverage.

Many visitors came to the Red Barn just to watch the construction of the Great Gallery. Staff took the opportunity to answer questions about what was being built, the schedule, and what would be inside. Accommodating this curiosity was one advantage to the difficult work of running full operations on a job site.

Construction was comparatively smooth for a one-off project of this kind. There were a few setbacks, but nothing that crushed either budget or schedule. The contractor gave credit to the high quality of the architectural and engineering team, an experienced group with accomplishments in historic preservation, public buildings, master site planning, and space frame engineering. Ibsen Nelsen and Associates surrounded themselves with highly experienced consultants for the project. Jack Christiansen, with Skilling, Helle, Christiansen, Robertson, was a talented partner of one of the largest structural engineering firms in the Northwest. The Skilling firm had designed the World Trade Center in New York and had experience with clear-span structures for public assembly.

Other design team members were carefully selected for their expertise and experience, including landscape architect Richard Haag of Richard Haag Associates; mechanical engineer Arthur L. Zigas of Syska & Hennessy; electrical engineer Gerald Fitzmaurice of Beverly A. Travis and Associates; and energy systems consultant Vladimir Bazjanac, Ph.D.

With the foundation complete, the Great Gallery could now start to take shape.

Specialized consultants were added for tasks in lighting, theater equipment, and exhibit design and production. Such a cadre of specialists had seldom been assembled for a regional museum, setting the bar high for the MOF in its expansions over time.

While the expansion was under construction, the museum was open every day. Visitors continued to stream through the doors, educational programs were conducted as usual in the Red Barn basement, and events were produced. The operations and capital campaign staff hunkered down in the Phase 1 structures and ran a first-class museum and a triumphant campaign. Throughout the years the docent corps had doubled from 15 to 30, and with the realization that the opening of the Great Gallery would require even more, a call for docents was sent out in early 1987. After an overwhelming response, 52 more docents received training and were certified in March 1987.

Much attention was focused on the new construction. The huge clear-span building was designed to accommodate the hanging of tons of valuable artifacts in a simulated flight environment. Nothing of this scale had ever been done in the museum world, locally or nationally. The engineering solution was simple, at the same time sophisticated. The strong triangulated space frame provided vast clear spans of space for large artifacts. The strength in the steel frame welcomed hanging points for tons of weight along the upper and mid levels of the space. These features provided a flexibility within which exhibits and displays could be changed out often and with comparative ease.

Features such as the huge doors engineered into the southern elevation made the gallery eminently accessible, tolerating many artifacts without disassembly. The basement-level connectors formed a support system back-of-house for catering special events. Commercial-grade elevators at several points handled heavier loads for provisioning banquet lounges. The early planning based on requirements in the Blue Book led the way to a flexible, functional building that just happened to be beautiful. As architect Nelsen summed it up: "This building should be of the same quality as the stuff they make across the street."

The 186,000-square-foot museum was completed in 1987 for a total cost of $33.2 million, which included the first phase of the Red Barn, the Great Gallery, and all the attendant costs for exhibits, furnishings, equipment, and financing. The facility was bigger in size and better in quality than planned. The construction cost of $121 per square foot and the total development cost of $178 per square foot (all project costs for administration, design, construction, furnishing, equipment, exhibits, taxes, financing, and contingencies) were low to moderate when compared to other contemporary museum projects.

For example, the downtown Seattle Art Museum was built a few years later at a total development cost of $335 per square foot. The Virginia Air & Space Center, the

Top: Hanging the strengthened steel frame for the glass roof.

Bottom: Great Gallery completion ceremony. Left to right: Ibsen Nelsen, Don Moen, and Howard Lovering, 1987.

first flight museum constructed in the United States after the MOF, cost $270 per square foot. The latest flight and technology wing opened at the Franklin Institute in Philadelphia during that period at a cost of $40 million for 100,000 square feet, or $400 per square foot. In 1988, when the Smithsonian added its new glass-and-steel restaurant to NASM, a structure sometimes likened to the Museum of Flight glass gallery, the construction alone cost $367 per square foot. The dramatic, unusual, and beautiful museum was accomplished on a moderate budget.

In anticipation of the museum's grand opening, The Boeing Company offered to underwrite a celebration film. Gary McMillen, a Boeing employee and frequent contributor to the museum, produced and organized the project.

"This building should be of the same quality as the stuff they make across the street."
— Ibsen Nelsen

Great Gallery Campaign

The Great Gallery campaign was the mother of all campaigns at the MOF. Those who endured the long process recall it as setting the base for all fundraising to follow. John Hembroff, the campaign director, would go on to manage one of the largest campaigns in Northwest history, but he said of this initial MOF task: "Of all the fundraising I have been involved in over 42 years in the business, the MOF experience is most memorable; it meant a lot to all of us." Alison Bailey worked with Hembroff on that campaign and recalled: "We met many challenges and learned from all of them. So much was new, and enough of what we did worked to accomplish more than most thought could be done. All of us were veterans after that experience and felt we could succeed at any campaign that came along, we could deal with anything thrown at us."

T. A. Wilson (left) shows Governor John Spellman a model of the Great Gallery, 1984.

Funding for Phase 2, with a target of $26.5 million, took several years and a lot of hard work, although it did have a base upon which to build. When T. Wilson accepted the leadership, it made the campaign feasible. He organized a corporate support group to make sure the campaign was successful. John J. Italiane, a Boeing executive, was assigned full-time, working directly for Campaign Chair T. Wilson. Italiane took the role seriously, making sure, as Boeing managers are trained to do, that there were targets and schedules to be addressed and to be met.

John Hembroff was recruited to be the first museum staff development director. Hembroff was employed as director of development at the Seattle Art Museum (SAM) when he met Lovering on a mayor's task force to look into prospects for the Georgetown Steam Plant. Lovering informed Hembroff of a new position at the Museum of Flight and invited him to interview. Hembroff, an Air Force veteran, was intrigued, although he later admitted he thought the campaign risky. He interviewed and was selected.

In turn, Hembroff brought another senior staffer from SAM, Randa Cleaves, and began to fill the few positions at the MOF with experienced people he knew. He called in Alison Bailey, an intern at SAM, to offer her a position, warning that it was temporary. Bailey recalled that she liked what she saw and said to herself, "This may be temporary, but for me it will be permanent." Fortunately for the museum, she is still there and working wonders.

Hembroff and staff were not only good at what they did, but for the first time there was fundraising staff that worked as a team with operations. Lovering recalled, "We were a good team with mutual respect, and that made a difference. We all had the same objective."

The involvement of T. Wilson meant more than any promotional strategy. It answered questions that suppliers and subcontractors might have had and made clear that the company was behind this project full force. With John Italiane leading the effort, the results came across from all parts of the country. The continued and increased support of the Boeing employees was particularly gratifying. Even with some union and association disputes with the company, the rank and file stepped up in additional solicitations to pledge another $6 million before the campaign was over. In total, individual employees contributed more cash than did the corporation, well beyond what any of the fundraising advisers had ever projected. All parts of the community and the aerospace world of suppliers could see that this was a project gaining great momentum and worth the investment.

The campaign attracted more out-of-state contributions than any other cultural institution at the time. This was the result of the aerospace reach into the international supplier and customer base. The combination of community and corporate leadership, joined with tremendous and unexpected support from the employee base, produced a successful campaign to build out the Great Gallery. T. Wilson was honored with his name on this key structure. John Italiane and John Hembroff were also saluted for their efforts. Hundreds who participated in solicitation and support enjoyed the pride of being part of one of the most successful campaigns in Northwest history.

The Great Gallery campaign and 1980s staff reunite at the 50th Gala. Alison Bailey, Randa Cleaves, John Hembroff, Jane Brammer, Howard Lovering, Peggy Nuetzel, and Cindy MacKenzie.

The idea was to create a dramatic film about flight, from dreams to real accomplishments, incorporating regional history and featuring major artifacts in the museum collection. The film would require a substantial budget for a high-quality production, and the decision was made to have it do double duty as a visitor-orientation film for the museum theater.

What was not resolved was the perfect narrator. The team discussed, evaluated, and listed candidates, but there was only one in the top position, Walter Cronkite. Everyone knew this was a stretch, but it was the perfect idea.

The budget was not sufficient to attract someone of his stature, but Community Relations Manager Cindy MacKenzie took on the mission, called Mr. Cronkite's office, and reached his assistant. The answer, though polite, was no. She reported that Mr. Cronkite was too busy and that he now seldom traveled. MacKenzie called again, making a friendly connection with the assistant and offering Mr. Cronkite a ride on a B-17 as an inducement. Cronkite's assistant granted that this was attractive, but the answer was still no. It appeared that there was no way to get the person everyone wanted.

Then one afternoon, MacKenzie excitedly announced that the assistant, citing the fact that she admired MacKenzie's tenacity as much as the significance of the offer, said that Mr. Cronkite would not only narrate but

Top: Cranes hoist the Douglas DC-3 into position in the Great Gallery, 1987.

Bottom: Who better to help celebrate the gallery opening than Mercury 7 astronauts?

volunteer his services if the film crew would travel to his offices and studio in New York. McMillen, MacKenzie, and the crew were dispatched to Cronkite's studio at CBS offices in Manhattan. Mr. Cronkite approved the script and, with his elegant voice, narrated the piece in one take.

The preparations for the elaborate Grand Opening were made with some good backing and luck. Staff agreed that it would be phenomenal to have the Mercury 7 astronauts as featured guests. These first-in-space heroes had established a charitable foundation, and their agent was contacted.

A deal was proposed that for a fee of $25,000, for use in its college scholarship program, the foundation would send representatives to the Grand Opening. Rainier Bank made a pledge of $25,000, assuring their attendance.

Lovering had envisioned a transportation museum so beautiful that a symphony orchestra would want to perform in the space, and the Bellevue Philharmonic eagerly accepted the invitation to play for the grand opening banquet.

The museum invited President Ronald Reagan, and they were not surprised to hear that he could not attend. Of course, their approach had been to start at the top—that was what this new institution deserved.

Trustees Bill Boeing Jr. and Wells McCurdy next invited Vice President George H. W. Bush. The Vice President had many fans in the region, had been a WWII combat pilot, and was a champion of youth education. The board was pleased to hear that the Vice President would attend.

The Secret Service and the advance team made all the arrangements, and the Vice President was a delight from the moment he deplaned from Air Force Two at Boeing Field. The Vice President and his entourage toured the new museum before the ceremony, allowing the honored guest to sit in his favorite aircraft for photos. A phalanx of media followed the group, and the footage was broadcast for weeks to come. The Vice President clearly loved the museum, the aircraft, and the exhibits, and was impressed with the programs for motivating and educating children.

The festivities began on Friday, July 10, 1987, with an opening dedication and ribbon-cutting ceremony on the museum grounds. There were hundreds of attendees, young and old, at the invitation-only event. The master of ceremonies was Jack Link, radio announcer, media executive, and aviation enthusiast. Bill Boeing Jr. recruited Link as emcee as well as for other museum chores.

Vice President George H. W. Bush (center), with Dick Bangert (left) and T. Wilson, helps cut the ribbon for the Great Gallery opening in 1987.

Richard E. Bangert, chairman of the Museum of Flight Foundation, welcomed the crowd and spent a few minutes tracking the two decades of work that had set the stage for this important day.

He introduced T. A. Wilson, chairman of The Boeing Company and also chair of the capital campaign that had successfully funded this major step in completing the initial museum complex. T.'s remarks were respectful but short, with no excesses. "I have to admit it's a glorious facility," said Wilson, who described himself as an early doubter won over by the enthusiasm of Boeing employees. "It's one hell of an asset to the community."

In turn, Wilson introduced the Vice President of the United States, George H. W. Bush.

Mr. Bush delivered a short and spirited welcome. With a look of sincere delight, he spoke of his early fascination with aviation, his entry into flight training, and some of his military service, always with a light touch. He remarked on the museum's great potential to capture the attention of children and to motivate them to respect learning, history, and technology, even channel them into aerospace careers. "Think of the sense of wonder this museum will give the youngsters," said Bush. "They'll see the incredible progress we've made since Kitty Hawk. We're improving air travel so rapidly that, imagine, people could walk into the museum in a few years and consider the 747 may be a relic, a thing of the past."

It was a memorable presentation at a museum dedicated to education, and to those in the audience who had worked so long and hard, it was a sensational christening of this new institution.

Other speakers were the representatives from the Mercury Seven Foundation, including Commander Scott Carpenter, Betty Grissom, widow of Gus Grissom, Captain Walter "Wally" Schirra, Admiral Alan Shepard, and Major Donald "Deke" Slayton. Shepard spoke of the new museum as a place that can motivate the public to encourage space discovery. "I worry that we're somehow getting complacent today," said Shepard, who on May 5, 1961, became the first American in space when his capsule *Freedom 7* was rocketed 116 miles high on a 15-minute ride.

The first Great Gallery exhibit was an impressive backdrop for the opening celebration.

"This reminds us of yesterday's heroes. It challenges us by what was done yesterday and what needs to be done tomorrow."

The Mercury stars enhanced all the events of those three days. They were personable and articulate representatives of the early space program and were a colorful addition to the grand opening. The U.S. Army Ninth Infantry Division Band provided the music. The presentation, though lively and memorable, was short and sweet, much in line with the preference of T. Wilson.

Mercury 7 astronauts at the opening of the Great Gallery. From left to right: Scott Carpenter, Donald "Deke" Slayton, Alan Shepard (speaking), Betty Grissom (widow of Gus Grissom), Walter Schirra.

At the luncheon in the Red Barn following the ribbon cutting, Richard Bangert presented the prestigious Chairman's Award to members of the design and build team. This award, previously conferred on Howard Lovering for his management and John Hembroff for guiding the capital campaign, was now bestowed upon architect Ibsen Nelsen, general contractor Don Moen, and museum owner representative Kent Kistler.

Chair Bangert's awards and comments aptly portrayed the cooperative spirit of the museum team and all who participated in its construction. "Ibsen Nelsen has served our Foundation for as long as I have, some 13 years, first as a volunteer. Ibsen and his people, particularly Lidija and Ivo Gregov, project architects, took on this project as if it were the most important job in the world, which of course we thought it was."

Bangert praised the commitment of the designers over the years and their sense of balance with the old and new: "Consider how they blend together on this airfield site and you understand why critics are paying respects to the design."

Saluting Don Moen of Gilbert H. Moen Company for his calm and friendly demeanor while supervising the project, which was completed on budget, Bangert referenced the contractor's character: "Don started this job in a curious way by making a sizable cash donation to the campaign. I wondered how long it would take for him to try to get it back with change orders. I am still waiting."

The museum selected staff member Kistler as the owner's representative from his position of assistant curator. He had previous airline experience with facilities and was a good fit. With the award, Bangert commented: "Kent took on the assignment with a vengeance. He was a consummate team player with the architect and builder. His steady manner made all of us more comfortable during the building process."

The Boeing 80A, which marked the beginning of the museum itself, was front-and-center in the Great Gallery opening.

Later that evening, guests, dressed in tuxedos and gowns, slowly made their way to the east entry of the new museum. It was a precious and promising night. Those privileged to attend—community leaders, major donors, and longtime supporters—relished this occasion. Attending grand openings and fine banquets was nothing new to most in attendance, but there was an aura about this evening and this magnificent new institution, and the promise it was bringing to the community.

It was a warm night, with clear blue skies, and the smell of sweet July air pushed out the occasional bursts of aviation fuel. Operations in and out of Boeing Field served as a colorful divertissement for the theatrical production of the evening to come. The row of lights in the setting sun illuminated the east side of the airfield and the backdrop of Beacon Hill. The museum's incandescent glow spotlighted aircraft suspended in the Great Gallery.

This introduction to the site was as dramatic to those who had not followed the early operations of the museum, or the construction of this massive expansion, as it was to others who had watched or spent decades working on this evening.

The museum was enchanting in the late sunset, with the changing light falling in varying shapes around the huge space and its artifacts. A palpable expression of awe soared within the crowd of revelers, and the ambient noise rose to a noticeable din. Over cocktails, friends shared stories of the early days, of the aircraft, or local history. The call to dinner was late, as could be expected, and the conviviality moved from the bars and exhibits to the side gallery tables set for the dinner. Westin Hotel, with its highest level of catering, used warming and cooling carts in the back of the house, as the catering kitchen structure was not yet built. The service was smooth and flawless, the meal excellent.

The program's introduction and remarks were short, saluting the successful fund drive and those who contributed to this fine institution. The Great Gallery was equipped with a drop-down screen of theater proportions. The first film, *Building Complete*, produced by The Boeing Company, was a short, instructive, and visually appealing story about planning and building the Museum of Flight.

The centerpiece of the evening was the film *The Dream*, which used the museum's collection to trace, with Mr. Cronkite's marvelous voice, the evolution from early flight to the jet age and beyond into space. At the film's end, with a blast-off of Apollo to the Moon, sound effects rocked the Great Gallery.

Then, out through a release of faux rocket vapor stepped four of the original Mercury-Gemini-Apollo astronauts—Alan Shepard, Scott Carpenter, Wally Schirra, and Deke Slayton—as well as Gus Grissom's widow, Betty. This evening and the remarks of the celebrities set the tone for the future of the Museum of Flight.

The next day was the premiere for donors and members, followed on Sunday, July 12, with the opening for the general public. It was the start of the new era of proving the value of this institution, but for now it was all about joy and pride and a real sense of a job well done.

"The Dream" Takes the Gold

In a scene from "The Dream," a Lillienthal gilder is readied for flight.

Top: A scene from the museum's film *The Dream*, narrated by Walter Cronkite, which won a gold medal at the International Film and TV Festival of New York.

Bottom: With its glass walls and ceiling and sophisticated lighting, the Great Gallery is a breathtaking space in the evening.

Thousands flocked to view the brand-new facility and marvel at its size, with lighting and shadows shifting and changing color with the daylight admitted through the glass. This new gallery, though immense in size, proved perfect for the display of large artifacts at several levels. With its reflective, crystal quality, it did not dominate the Red Barn, seeming always to keep the historic structure in a place of eminence. Even though the exhibit investment was modest for a large new museum, the beauty and comfort of the gallery environment satisfied visitors.

Few facilities of this magnitude had been built in the region and none like this massive but beautiful clear-span space clad in glass. Thousands of visitors each year could watch and just imagine what was to come within this remarkable envelope. Expectations were raised. The museum's response was to ramp up governance and operations to a high level and to set new standards for flight museums. That was done.

But just as consultants from the Smithsonian National Air and Space Museum had warned, after the euphoria of opening wears off, the cold reality of the daily running of an institution sets in. Everyone was tired yet now faced with the daily drill of visitor service. The facilities were grand, but the collection was thin and the interpretation in the formative stage. Eager volunteers needed organizing and training.

The board members, finishing one of the largest capital campaigns in regional history, were fatigued and not prepared for additional fundraising. The operation budget needed to be generated almost entirely from earned income.

But even during this early operating period, the team located other artifacts, increased public programming, won educational awards, organized events of broad popularity, and pulled off a few stunts that attracted international media coverage. The museum was on track in proving that its event-driven and educational model was viable.

Lovering remembered a cold moment when he realized the need for operating cash. The capital campaign had not provided a reserve. There were bills to pay and a payroll to meet. Money was needed to stock the gift shop, immediately popular even with minimal inventory. The museum needed a line of credit to shore up cash-flow issues. Lovering was introduced to the Bank of Tokyo, which expressed an interest in helping. This connection was made through the Paper Airplane Contest and other partnerships in Japan. The Bank of Tokyo graciously offered a line of credit at a favorable rate and required no collateral.

The board approved and the deal was done, signed off on a minimal form similar to a car loan. The museum operations now had a backup, and over the next year only had to make one draw-down of approximately $60,000 to meet payroll. Fortunately, gate income almost immediately caught up with costs, the loan was paid back, and the line of credit was not needed again. The lesson learned was that any new institution needs an operating reserve when it first opens. With this lesson in hand, the museum was off and running full size and full force.

Red Barn and Great Gallery at night.

Museum Design

Artist's concept of the Great Gallery, 1985.

One of the most interesting features of the Museum of Flight has been created from a dilapidated building parked at the southwest margin of Boeing Field. Future generations will thank the Museum of Flight Foundation for saving "The Red Barn." It represents an important part of our region's aviation history. It is a landmark building of national importance. It is a scarce and fine example of a turn-of-the-last-century wood-framed industrial building. It is the building in which one of the most sophisticated aircraft companies in the world was started. Boeing is known internationally, and this modest building, such a contrast to the perfection of the company's modern aircraft, will become familiar in future years to hundreds of thousands who will visit the museum. How many great industries today can point to the building in which it all began?

As a structure the Red Barn is a pure example of American wood-frame architecture. While this type of industrial building was once common, only a few examples remain. The wood frame, with which many of us are familiar in our homes today, was not always the basic structure in American wood architecture. It was invented in Chicago, according to the historian Sigfried Giedion, in 1820 by George Washington Snow, a builder who owned a lumber yard. He saw it as a cheaper way to put up a house than the post-and-beam system it eventually replaced. The simple continuous bands of wood windows, the repetitive wood frame structure exposed on the interior, and its shape and scale give the Red Barn its characterizing identity. When it is fully complete, with its exhibits, this handsome building will become familiar world-wide as a symbol of early aviation, and as the birthplace of one of the great aviation companies of the 20th century.

A surprising number of people are still around who worked in the Red Barn in its earliest years. They and a lot of other people have great sentimental attachment to the building. It was built originally with parsimonious economy; you could say it was a cheap building. There is only one layer of boards on the exterior, one-by-six-inch horizontally applied boards with a crack between every pair. It's not hard to imagine that the workers found it a bit chilly in winter with so little separation from the weather. Phase 1 of the museum involved placing the Red Barn on its permanent foundation and completing the restoration. The subsequent phase to follow will house the museum's valuable collection of historic aircraft. It will be an unadorned structure of concrete with prefinished industrial metal roofing, and is designed to complement the Red Barn.

The design of the new building, its configuration, grew directly from the museum's program and from the requirements of the site. The dominant feature of the building will be the great hall of aviation, in which can be suspended and displayed historic aircraft such as the 1916 Boeing B&W replica, the Grumman Wildcat, the Lockheed TV-1 Shooting Star, and the venerable B-17 Flying Fortress. This dramatic setting is conceived as a transparent and liberating space, a suitable environment for these splendid aircraft. It was important that the B-17 be displayed indoors; the large gallery will enhance the Fortress yet keep it from dominating the museum.

The defining parameter of the site is the slope-clearance requirement from the edge of the runway. The building's roof slope conforms to this clearance line. The lower part of the building is devoted to smaller galleries, museum offices, theater, classrooms, archives, and the like. This relatively more enclosed structure acts as a foil, and will anchor and stabilize the great space-frame roof of the large gallery. There will be continuous decks along the upper floor, from which there will be an unimpeded view of the endless traffic of one of America's busiest airfields.

More than anything, however, this museum will bring to life some of the richness of our aviation history. It will honor those who created it, and communicate something of the energy and greatness of our industrial heritage.

— Ibsen Nelsen

Left: A view down through the ceiling of the Great Gallery and the multiple levels of exhibits it houses.

Right: The Hangar, added soon after the opening of the Great Gallery, was a hands-on educational experience for kids.

CHAPTER TWELVE

Keep Climbing

With the Red Barn and Great Gallery open, operations ramped up. The museum transitioned to a more structured organization. New hires came on board in a more systematic manner, several with extensive museum experience and education credentials, including a few from the prestigious Smithsonian National Air and Space Museum.

"The Hangar" exhibit for young people, added soon after the grand opening, was a natural evolution of earlier hands-on classroom programs. It was simple, colorful, and reduced in scale for young people to feel comfortable. It incorporated items to explain principles of flight in a playful manner, with games, puzzles, and dressing up.

The simulated aircraft hangar contained real, once flown small-scale aircraft, including a biplane, a helicopter, and a jet, actually a BD-5A built from a kit. All of these aircraft could be entered, which gave a sense of drama to the various lessons. Nothing in the exhibit was digital or electronic, all was mechanical, and docents enhanced the experience with stories, games, and exercises. In 1989, the FAA recognized The Hangar as the best aviation youth program in the nation.

Additional educators with expertise in teaching in the museum environment came on board, bringing respect to the institution. In 1991, Dr. Virginia Wagner, manager of education, was accorded the prestigious Nancy Hanks Memorial Award for Professional Excellence from the American Association of Museums, the only time this award has been given to a museum staff member in the Northwest.

Now with a Great Gallery to display large artifacts, the museum organization began immediately to hear of aircraft that might be available, some as donations from owners or estates. The board, staff, and volunteers would scan for opportunities and forward targets to the collection committee. MOF was now in the running for the best pieces available and competing for those that were on the priority list.

Bill Phillips was among the first contributors, generously donating his Stearman PT-17. This was not a rare airplane, but these trainers were relevant to what became Boeing Wichita Division and had played a role in pilot training for aerial combat. Phillips' aircraft, beautifully restored to flying condition, was a valuable addition and helped set the stage for new contributions to come.

Lovering joined pilot Phillips on one of the plane's final round-trip flights, this one to Yakima, Washington, and back to Boeing Field on a beautiful early summer day. As they approached the Yakima Airport, Lovering, in tribute to Captain Jack Leffler, donned a gorilla head for the landing. This trainer is still on display in the Great Gallery and speaks to the early generosity of private donors.

It is difficult for any institution to put away enough money in an acquisition fund to compete directly for an essential artifact. Therefore a collecting museum such as the Museum of Flight must establish a vital and sustaining collecting strategy. It takes some daring and some doing.

When an aviation star became available, the museum would develop a proposal and funding plan. A potential new artifact was quickly worked through the appropriate committee with a recommendation for action. If the acquisition was a go, a team was assembled to make a request, initially trolling for a donation or at least a suitable price. The museum's proven public service undoubtedly helped in these negotiations. If no adequate funding was readily available, typically the case, the team would attempt to buy time to raise the dough required. All of these dodges were used successfully, and this approach worked in acquiring the rare Ryan M-1 as well as the proto-type Lear Fan, proving a good strategy over the years.

In 1987, a true star presented itself in what seemed an unexpected and fortuitous fashion, but was the result of years of work. Trustee Dick Taylor had recruited Bob Bogash, an engineer at The Boeing Company, to work on the acquisition committee. Bogash would place a plane on his personal priority list, then doggedly track it. The first 727 was an early target. Entering service with United in 1964, it flew revenue passengers for 26 years. Bogash made pitches to United Airlines CEO Eddie Carlson and his successor Dick Ferris for the 727 long before it was surplus to need, much as he did with other priority targets.

In 1988, UAL CEO Stephen Wolf was in town with other company executives promoting his airline in the renewed competition for the Seattle-to-Tokyo route. Bidding with Continental and American, UAL pushed its brand in a series of events, some at the MOF, including flights for students in the historic 727. This was the chance for follow-through.

In a short conversation, Lovering suggested that an announcement of the gift of the first 727 could get media and community attention. Wolf and his entourage

immediately put the idea into the system. As the aircraft was then nearing the end of its service cycle, within days UAL announced the gift. It was an opportunity that was seized quickly when the circumstances were just right, but also was the result of early action and years of determined work.

In January 1991, the aircraft, painted in original United colors, made its last UAL flight to Boeing Field for a museum ceremony, then up to Paine Field in Everett, where it would be restored. No one at the museum or UAL could have imagined that it would be another 25 years before the valuable artifact was restored and flown from Paine Field into the museum on its last flight. Years of detailed work was accomplished by a team of volunteers under the leadership of Bob Bogash. The important decision to acquire this first of its kind set a pattern. The commercial aviation segment of the collection, begun with the 80A and Boeing 247, was adding some jewels.

Top: The first 727 at the Museum of Flight in 1988 for a United Airlines event. UAL announced that it would be gifted to the museum shortly thereafter.

Bottom: One of the museum's several eye-catching billboards produced by Livingston & Company, which won a 1992 OBIE Award for excellence in advertising.

It was a magical moment when Lovering was contacted in 1989 with the offer from Boeing of the first 747, *The City of Everett*. Vice President George Bush's prophetic remark at the 1987 Great Gallery Grand Opening that "someday one can imagine a 747 displayed at this museum" was proving real within just a few years. Despite the great obligation involved in taking care of these big commercial aircraft, it was an incredible opportunity to have them in the collection. They were grabbed when available, and the long-range plan for their restoration and display was worked into the system.

With his early death in 1990, Robert "Swage" Richardson, who preferred flying to museum prowling, left the museum one of the most valuable pieces in its collection. His magnificent B-17F was transferred at his original purchase price to the museum, along with the gifts of a Howard DGA and a Grumman Goose that could be sold to finance the acquisition. This was another generous gift that a supporter and board member made possible.

Friendliness and cleanliness were also in the museum's focus. Friendliness cost nothing, and

Top: Accession ceremony for the first 747, *The City of Everett*, March 28, 1990.

Bottom: Swage Richardson's B-17F arrives at the museum in 1990.

everyone at the museum was prepared to do what was necessary to make a visitor comfortable, even to help find a hotel, give directions to a restaurant, or find information on building a model airplane. All staff, whether in the office or working the floor, were ready for cleanup of any part of the museum. Though this created the near mutiny of some professionals, it was ultimately accepted as a general objective of operations.

Ralph Johnston took on the management of Visitor Services (his title) after the grand opening. He had graduated in nonprofit management at Yale, then worked part-time at the Metropolitan Museum of Art and later full-time at the National Air and Space Museum as theater manager. Ralph was taken with the new focus on visitor services in the big museums on the East Coast. Reshaping an operating model that had long favored the collection and its curators, museums began to prioritize the visitor experience, comfort, safety, and entertainment along with education. It was new and experimental, an organizational policy that assisted in making the MOF an attractive and entertaining place to visit. Throughout the past 30 years, visitor-oriented management has continued to be a mainstay subject of museum workshops.

Special events that fueled both participation and revenues were expanded. Every production had the objective of earning income, attracting new members, and promoting the museum. The museum always devised a way to receive national coverage. Each included an effort to provide something of interest for old and young, the flight novice and seasoned enthusiast, family and tourist. The media and community came to expect unusual and dramatic events. MOF never followed a traditional transportation museum profile, but rather imagined what might be done and took chances. The museum's event-driven model proved quite effective.

Specialized programs were enhanced with less frequent but much more dramatic events. Throughout the years they included the "Friendship One Around the World for Kids" record-setting flight and the "Apollo-Soyuz" anniversary reunion, both establishing the MOF as a prime location to rally aviation and space pioneers from this and other nations.

Friendship One was an outstanding event in January 1988 that set a round-the-world record, raised $500,000 for children's charities and museum education programs, and bonded 141 crew and passengers into an exclusive group with forever memories.

Friends Bruce McCaw, Joe Clark, and Clay Lacy, now all MOF

Friendship One passengers and crew before boarding the 747-SP.

Pathfinders, conceived the unusual project in conversation at the Paris Air Show. When they approached Eddie Carlson, chair emeritus of UAL, for his support he replied, "This is a heck of an idea! Let's do it." McCaw gave Carlson's enthusiasm the ultimate credit, saying, "We couldn't have done it without Eddie."

McCaw, Clark, and Lacy formed Friendship Foundation and began their search for supporters, which in addition to UAL included Pratt & Whitney and The Boeing Company. Volkswagen of America donated a 1988 VW Jetta (which means *jet stream* in German) to the Friendship Foundation. The "fastest Jetta in the world" was later gifted to the Boys and Girls Clubs of Washington for auction. Also in the belly of the 747-SP was a 10-foot model of the aircraft to be signed by passengers and crew upon completion of the flight.

With their friend Eddie Carlson's support and the loan of a 747-SP from UAL, the three longtime museum trustees and supporters detailed the logistics for a record flight and organized the experienced volunteer crew to fly and service the aircraft. The plan was to travel from west to east in an attempt to break the 1985 circumnavigation record of 45 hours, 32 minutes.

The fare for the flight was $5,000 per person, and all 100 seats were sold. Those taking the trip included aerospace celebrities such as Neil Armstrong, Eddie Carlson, Bob Hoover, and Moya Lear. Bob Mucklestone, an MOF trustee who in his Cessna 210 had set the solo round-the-world record for elapsed time in May 1985, and his wife, Megan Kruse, were along for the ride, as well as showman and auctioneer Dick Friel.

According to an Emmett Watson article in the *Seattle Times*, prior to the flight a young TV reporter unwittingly approached passenger Captain Jack Leffler to ask if he had ever previously flown in an airplane. Captain Jack replied, "Yes, this is my second flight. I just passed my Fear of Flying course with a test trip to Portland. Then they said I could try a longer flight. So I'm on this one." At the time Captain Jack had 32,000 hours in his logbook.

USAF communication specialists, docents, and trained students track Friendship One at the museum's Mission Control.

At 7:14 p.m. on January 28, 1988, the aircraft departed to circumnavigate the globe. After circling Boeing Field for a salute to the cheering sendoff crowd, the aircraft officially started its recorded flight at 7:27 p.m.

Enthusiastic students staffed Mission Control Center, set up in the Great Gallery and supported by the U.S. Air Force Communications Agency's "Hammer ACE" global satellite command from Scott Air Force Base, near Belleville, Illinois. This ground control tracked the flight progress with constant updates on airspeed and position and weather reports. Retired test pilot Lew Wallick answered questions on the history and background of the 747, and McChord Air Force Base personnel explained aircraft navigation and refueling. Mission Control Center provided tremendous worldwide exposure for the museum.

During the flight, passengers rotated between first- and business-class seats, dined on gourmet meals, played games, and watched films including *Top Gun*, for which Captain Clay Lacy had directed the aerial photography. Moya Lear, the widow of the founder of Lear Jet, pitched in to serve croissants over the Pacific. It was that kind of fun.

Fuel stops were in Athens, Greece, where the foundation delivered a donation to children's charities, and Taipei, Taiwan, which at 35 minutes, Clark described as being similar to a "pit stop at Indianapolis."

Flying in the jet stream at 33,000 feet or higher, and at an average speed of 629 mph, the Boeing 747 completed its 23,125-mile circumnavigation at 8:17 a.m. on January 30. Its recorded time was 36 hours, 54 minutes, and 15 seconds. Lacy had previously estimated that the flight would take 37 hours.

A crowd of more than 1,000 and the Auburn High School Marching Band welcomed the group of weary but jubilant travelers. Neil Armstrong remarked that his first steps on the Moon were "a walking kind of thing," but the record flight was a "flying kind of thing." Another passenger called the adventure the longest cocktail party of his life. Lacy was pleased that the 141 passengers were "still talking to each other." Jane Carlson Williams was overheard saying, "The last time this many people went around the world together, they were with Magellan."

Captain Lacy, an experienced United pilot, confided that this was the longest continuous flight of his career, as was most likely the case for all. With a total of 77 licensed pilots on board, he quipped that he had "more backups than a space capsule." The record chase, the camaraderie, the new friendships, the philanthropic purpose all added up to what is an enduring memory for participants and one of the remarkable partner events in the history of the museum. It was evident from the enthusiasm of the deplaning crew and passengers that this had been more than just a record-setter. It had been an experience of a lifetime.

The trip has been described in the memoirs of many of those on board and recalled at several anniversaries of the flight held at the Museum of Flight. At those gatherings, alumni of Friendship One continue to share stories, and there is always another financial contribution

Left to right: Joe Clark, Clay Lacy, and Bruce McCaw at the installation of the plaque celebrating the flight of Friendship One.

made to education programs at the museum. Friendship One lives on as only the best of events can expect to and proves again that connections such as those of McCaw, Clark, and Lacy can generate ideas to energize any institution.

This ability to produce high-visibility special events was not lost on the officials at the airport. In 1988, King County International Airport approached the museum to sponsor an air show that could promote airport facilities on Boeing Field.

Boeing B-17 Flying Fortress

No airplane is more famous than the legendary Boeing B-17. A four-engine heavy bomber with a crew of 10, it bore the brunt of U.S. bombing operations over Europe during World War II and was key to Allied success. So rugged was the Flying Fortress, and so dramatic its ability to return to base despite jaw-dropping battle damage, that it has achieved truly iconic status.

A restored B-17, minus gun turrets, flies over Mount Rainier.

B-17s of the U.S. Army Air Forces (USAAF) fought from the time of Pearl Harbor till the end of hostilities in Europe. They flew in every theater of the war, but it was their missions from English bases against Hitler's "Fortress Europe" that earned them lasting fame. At its peak in 1944, the Eighth Air Force had no fewer than 27 bombardment groups equipped with the B-17, each with four squadrons of 18 or more airplanes.

The Fifteenth Air Force had six more Fortress groups that attacked, in Winston Churchill's famous words, the "soft underbelly of Europe" from bases in Italy. "Without the B-17," said General Carl "Tooey" Spaatz, who commanded USAAF strategic air operations in Europe, "we may have lost the war."

The Eighth Air Force began humbly in 1942 with a handful of B-17s and an unproven philosophy. Instead of attacking targets near the front, it would fly far behind the lines to knock out factories and other targets, thus denying the enemy the means to wage war. Known as "daylight strategic bombardment," this unproved concept in turn relied on the secret Norden bombsight, a sophisticated gyro-mounted electro-optical computer allowing accurate bombing from high altitude.

U.S. Army Air Corps visionaries espoused the strategic bombardment doctrine in the early 1930s, but among the nation's aircraft manufacturers only Boeing thought big enough to give them the airplane they needed. Rolled out in July 1935, the Boeing Model 299—then the world's largest land-based plane—bristled with guns, leading a Seattle reporter to call it a "flying fortress." The name stuck.

Built in small numbers, prewar B-17s featured small tails, were underpowered, and lacked adequate armament. Early use of 20 B-17Cs by Britain's Royal Air Force revealed these and a host of other shortcomings just in time for Boeing engineers to perform a sweeping redesign before the United States entered the war. Despite these changes and subsequent improvements, initial B-17 penetrations deep into Germany experienced horrific losses to enemy fighters and flak. Schweinfurt, Regensburg, Münster, Oschersleben . . .

Air battles of mythic proportion raged over Europe. Eighth Air Force formations shot down hundreds of Luftwaffe fighters and were never once turned back by enemy action, but it was all at too high a cost. The lesson was clear: escort fighters were desperately needed with sufficient range to protect our bomber formations all the way to the target and back. With the arrival of North American P-51 Mustangs in early 1944, this need was met and the tide inexorably turned against the Luftwaffe.

Spearheaded by the B-17, the "Mighty Eighth" evolved into the largest aerial armada in history. Thousand-bomber raids became common, and some missions saw more than 2,000 bombers escorted by upwards of a thousand fighters. These large-scale operations helped bring Nazi Germany to its knees, but at a great toll. The Eighth Air Force alone suffered more than 47,000 casualties, with more than 26,000 dead. Seventeen medals of honor were awarded, many

Boeing Bee.

posthumously, to Eighth Air Force personnel. Of some 4,750 B-17s lost to enemy action, the vast majority flew with the Eighth.

The Museum of Flight's B-17, a rare F model called *Boeing Bee*, was built at Boeing Field just a mile north of its current home. Fully restored and flyable, this airplane lacks the operational markings worn by other surviving Fortresses. Instead it looks as it did when it rolled off the Boeing assembly line on February 13, 1943. This "factory fresh" paint scheme recognizes the contributions of the people of the Pacific Northwest to winning WWII.

The museum was wary of air shows for several reasons. First, the region already produced air shows at Paine Field, which was one of the largest in the Northwest, a well-established fly-in at Arlington, and one of the biggest military shows in the U.S. at McChord Air Force Base. Air shows are expensive productions with a lot of risks, and although popular, are lucky to break even in cost. They involve a mass of people and a lot of money over a year of planning, and it takes only a few days of rain or one accident to set back or ruin the proceedings. The museum would have demurred, but the County argued that this event would be good for the airport, and it was obvious to all that the museum owed a debt to King County.

The museum agreed to partner with the airport for an air show in July 1988, titled the Emerald City Flight Festival, which would bring out the best of the flying events and also promote the airfield. It did just that, with a goodly amount of flashy flying, in civil and military and vintage aircraft, but also a lot of static educational exhibits.

The museum contracted Metro buses to move crowds of people around both sides of the airfield. With trained docents positioned on all the buses, visitors were given the history of the airfield as they drove between venues, turning the air show into an interpretive experience for the airport. It worked.

The Soviet education delegation visits the museum's Lear Fan exhibit at Sea-Tac Airport with museum event manager John Ferguson (second from left), Dr. Valentin Shukshunov, astronaut Michael Collins, Dr. Yuri Karash, and others.

This air show became one of the event highlights of the summer season at the museum and for Seattle for a number of years. The museum became skilled at finding new themes and securing the best of the air-show acts. Sponsors were found, as is so necessary, in each of the years to make the event viable. Over those years themes saluted the military, 75th anniversary of The Boeing Company, and the state centennial. Hundreds of volunteers were trained under an air-show boss, Cyndi Upthegrove, who had the skills to manage the annual event.

The re-creation of the Apollo-Soyuz mission on its 15th anniversary and notably the famed "handshake in space" was a highlight for the Emerald City Flight Festival activities at Boeing Field in July 1990. The museum invited astronauts Donald "Deke" Slayton and Thomas Stafford to join cosmonauts Alexei Leonov and Valeri Kubasov, all of whom along with astronaut Vance Brand had participated in this first of the joint space flights between the United States and the Soviet Union in 1975. Leonov, the first to walk in space in 1965 and the Soyuz commander, said of the cooperation in space: "I believe our flight will be considered the beginning of broad cooperation between our countries. In the future, humanity will settle space and our mission will be a reminder of how it all began." The reunion was an instant success, bringing in larger than usual crowds to the Boeing Field air show and garnering continual media attention. All of the gentlemen knew one another well and relished the opportunity to be together and to commemorate their adventure.

In the re-creation ceremony staged at the air show, the four representatives spoke of their experience, which made the national news. Lt. General Tom Stafford commented that "the Apollo-Soyuz mission was the epitome of goodwill through flight—if you have a common goal you can work together to achieve fantastic results." Soyuz Commander Maj. General Alexei Leonov inspired the crowd with his heartfelt words in English and Russian: "We want to work together in the future. This air show speaks of the high minds and good will of all involved in making it possible— Soviet and American planes here side by side."

Top: Re-creation of the Apollo-Soyuz handshake on its 15th anniversary in 1990.

Bottom: A Sukhoi SU-27 Flanker guest-stars at Emerald City Flight Fest in 1990.

First reading a message from Vance Brand, unable to attend due to preparation for a Shuttle mission, Deke Slayton addressed the occasion, followed by cosmonaut Valeri Kubasov. Helen Jackson, the widow of Senator Henry M. Jackson and museum trustee, shared with the crowd that "Apollo-Soyuz was the harbinger of the thawing of the Cold War—our nations are moving closer together than they have ever been for generations." Closing out the short, poignant ceremony that featured the two countries' anthems and flags, Lovering concluded, "It's really a very simple thing we celebrate today, hands from the east and hands from the west clasped in friendship."

Though modest in presentation, this was another milestone event in building the museum's space themes and demonstrating what a great venue the museum could be to interpret and celebrate these achievements. During Flight Fest, this cooperative venture had also hosted the fly-in and display of the then-contemporary Sukhoi SU-27 Soviet fighters, a tremendous attraction to spectators in the first visit of these advanced aircraft to North America. The museum was reaching out internationally to form alliances and partnerships that would build up its image.

On July 11, 1990, the museum consummated its years of communications with the USSR and opened the first of its Soviet space displays, entitled "Conquering Cosmic Space."

In partnership with Eastern Washington University and its sister institution, Kalinin State University in the Soviet Union, the MOF signed a Protocol of Intentions to serve as a venue for this and a proposed series of joint exhibits. Washington State Governor Booth Gardner and museum officials greeted Dr. Valentin E. Shukshunov, Deputy Chair of the USSR State Committee for People's Education, and his Soviet delegation at the opening of the museum exhibit.

Emerald City Flight Fest was deemed a success for all but the museum. Its organization took a year, lots of staff, and a sizable budget. To assure success, all employees, docents, and volunteers had to work the event. Always having to worry about inclement weather or an accident, the team was exhausted at the end of the day. Even after a show that drew many thousands of satisfied visitors, the Monday after was somber.

Bart Hunt of The Boeing Company created the artwork for the Museum of Flight's first traveling exhibit, "Wings over the Pacific," in San Francisco.

Let's Do the Numbers

With the opening of the Great Gallery there was the opportunity to measure performance of the full-functioning museum. It is typical for museums to record their public service with headcount: how many annual visitors are attracted and how many users of the offered programs. This quantitative measure of service is often useful in applying for various grants in order to meet some program requirements for audience.

Another of the measurements is the number of members, a quantitative number that indicates community support. An attraction will also underscore the number of tourists—visitors from outside the immediate service region—in order to prove economic impact that might justify government support. Though instructive, these numbers can seem more important than they are. The result is that there can be overcounting and, even worse, overlooking what is really important to public service—the quality of the visitor experience.

Quality requires accounting along with quantity. The museum measured up well with both. The MOF enjoyed good quantitative metrics in the early years of the Great Gallery, with 424,000 visitors in the first full year of 1988. The internal business plan proposed 400,000 as a viable number to target in order to support the operations at the gate. That was attained.

Membership rose to 18,000 in those years, one of the largest numbers among all institutions in the Northwest. This number was somewhat inflated by the gathering of members during the large capital campaign, and not all of these would renew annually. But this was a large base, and it proved a loyal source of volunteerism and annual contributions. Membership is a great strength of the museum.

Early estimates of visitors from out of the region, including numbers from visitor surveys, proved that the museum was a tourist attraction. More than 50% of visitors in those years, and continuing, visited from outside the three-county central Puget Sound region. A large number were from foreign countries. This might be explained by the growing popularity of flight museums, particularly in a visit to an aerospace capital such as Seattle. There is no question that the MOF contributes a lot to the tourist economy, as well as to the profile of Seattle as a great place to visit and to learn.

Most important, what staff learned is that it is the quality of the visitor experience that proves value. The airfield is busy from the time a visitor steps out of the car or taxi until leaving. The aviation backdrop gives context to the experience, whether a casual tour of exhibits or a particular program. Visitors enjoy the atmosphere in this location at one of the busiest general-aviation airports in the country. The programming, exhibits, interpretation, symposia, presentations, and youth camps, special events, and overall education all benefit from the dynamic environment.

Air show acts are comparatively expensive, and costs are hard to control as the medium is so fluid and what is needed at the moment must be purchased. Within a month or so, as invoices were received, the costs were tallied. Every year, projected budgets were exceeded.

The airport was supported and promoted, hundreds of thousands of visitors were entertained and even educated, and the museum proved it could put on quite a show. Other than that, and that is substantial, there was little of sustaining value. The museum began to look at other types of events to promote its objectives and build net revenue for operations and educational programs.

In 1989, The Bank of Canton of California asked the museum to develop an exhibit on aviation in the Pacific Rim. The bank had refurbished the adjacent historic U.S. Subtreasury building at its headquarters site in downtown San Francisco as the Pacific Heritage Museum and wanted to partner on a temporary exhibit about how aviation had opened trade from the West Coast across the Pacific. With a budget from the bank and Boeing Company contributed services, the team prepared "Wings over the Pacific," which opened to the public on June 1, 1990. It ran for more than a year and was the museum's first traveling exhibit.

To celebrate the Washington State Centennial of 1989, organizers decided to produce and promote a number of regional events. The museum jumped at the opportunity and agreed to be part of a statewide salute to aerospace. Flight Fest that year served as a mother ship for the coordination of the centennial commemoration.

As the date approached, the museum found itself in a leadership role by default. MOF chartered the new Wings Over Washington Foundation, a nonprofit organization, and brought on a volunteer board that included MOF leadership to plan the centennial tribute to aerospace and to do so at a fast pace. Sponsors were lined up to put on an aviation event at Boeing Field and fly-overs at other venues.

Lovering remembered the tremendous energy that Concorde had produced on its visit in 1984. British Airways (BA) also had a good memory of the great success of the Concorde's first charter to Seattle. The Concorde was losing money each time it flew a regular route, even with what the casual traveler thought were high ticket prices. But it served as a brand-building component of the company, as it was still special and unique. It was pure fate that one member of the museum board, D. P. Van Blaricom, suggested a meeting with the regional director of BA. Van Blaricom and Lovering met with BA and came up with a partnership arrangement that would bring the Concorde in again at a reduced cost to the museum, and with a lot of fanfare for the airline.

The cost was substantial for any nonprofit, almost $1 million estimated for the flights and associated events, a third of the museum's annual budget. Although the wisdom of such a big investment was disputed, Lovering and Van Blaricom took the proposal to the board. Interest focused on the scale of the event and its contribution to the Centennial, but there was no enthusiasm for the financing. Finally, a go-ahead was authorized based on finding the funding.

British Airways
Concorde visits the
Museum of Flight for a
"Flight to Nowhere,"
July 29, 1989.

The promotion focused on passengers arriving or departing on flights of the Concorde while spending a week in London and environs. The museum hired an experienced travel adviser to put together what was hailed as "The Flight of the Century." It was two flights on the Concorde— one from London to Seattle and the other from Seattle back to London—and included one leg across the pond first class on a 747 jumbo jet.

The busy week in London was filled with special lunches and dinners and tours of historic sites, an evening at the theater to see Andrew Lloyd Webber's *Aspects of Love*, tours of British flight museums, a Thames riverboat cruise, and a day-tour on the famed Orient Express train. It was a terrific trip at a reasonable cost of $6,995, at the time not much more than a Concorde commercial flight from London to New York. The package also included Concorde presentations at the museum and two additional Flights to Nowhere—short flights to and from Boeing Field— while the aircraft was in Seattle.

The museum had some wiggle room for payment to BA, as per the agreement, the cost would not be billed until after the flight. Gary McMillen of The Boeing Company created an elegant brochure that was distributed to thousands of museum members. High hopes were that this well-planned event would take off immediately, but that was not to be. Weeks wound into months with a tepid response. Lovering appealed to everyone he knew. The tour and all it meant to the museum and the Centennial just did not pick up travelers. Early reservations were few.

Lovering resolved that the deal was struck and now somehow had to be implemented. There was no money to cover this mistake. Lovering's nights were sleepless, and he was eager to get to work early each morning to find out if there was some good news. It didn't happen for some time, almost too late. The museum needed 100 passengers each way to make a good net on the project, at least 60 to break even. These numbers now seemed insurmountable. The slow response was wearing down staff.

Late in the game, museum staff generated the interest of PEMCO's Stanley O. McNaughton, a museum trustee, local civic leader, and member of the King County Museum of Flight Authority. He signed up and reached out to his family and friends, and suddenly provided a large group of travelers. With this uptick in interest, things began to accelerate, and a week or so before the first leg of the flight, the museum had 60 confirmed reservations in each direction. Emboldened by a break-even, the event was plastered over the media. Nothing could now stop the Flight of the Century.

With few days to go until takeoff, there were approximately 20 seats unfilled. The deputy director suggested that the museum might consider letting staff fill those seats in return for the incredible effort they had performed running the museum and supporting these crazy ideas.

Lovering agreed and offered museum employees free seats on the Concorde if they took vacation time and paid for the other leg of the flight, and most importantly, if the paying passengers would unanimously agree. Many staff responded, and Lovering, who had scheduled pre-flight briefings as part of the fun of this adventure, brought up the subject with all paying travelers. The vote was unanimous and enthusiastic. Museum employees from curators to security guards flew the Concorde.

This fantastic event was just that. Governor Booth Gardner could not afford the time, but his wife, Jean, who had done a lot of the road work for the Centennial, was a delight on the Concorde journey, joining in at every lunch and museum tour as a gracious and engaging ambassador from Washington State. The trip to and in London was sensational.

Travelers had a great time but were eager for the flight on the Concorde back to the museum. They were to find that much of Seattle was also interested in the return. Lovering was called late at night for a radio station interview on the incoming schedule and the experiences on the trip. This was an indication of what might be ahead as a greeting. It was clear that the city was eagerly awaiting the Concord.

While in London, Lovering (or Lord Lovering, as his BA friends cleverly added to his ticket) was invited to visit Concorde headquarters for a tour and meeting. He learned of the great affection for this unique aircraft as a vanguard in the industry, and the enormous effort to keep the fleet maintained and in high-quality operation. It was obvious to Lovering that this was an aircraft of magnificent proportions for storytelling, one that would be a must-have for any museum.

During this tour and meeting with BA officials, Lovering reaffirmed the desire to receive one of the aircraft when and if ever declared surplus, and pointed out what a great home it would have at MOF considering its grand reception. All officials of BA seemed to agree that this would be a fine gesture. After this meeting, the Concorde brand manager, George Blundell-Pound, gifted Lovering with a book on the aircraft, with the encouraging inscription "Keep a place for us." It was apparent that there was no schedule for this acquisition process, estimated or otherwise, and that so much could easily change over the years. But MOF was in the running, though there would be stiff competition.

After a week of fun and games in London and environs, all the travelers at Heathrow met in the salubrious Concorde Lounge for cocktails before boarding Concorde for the memorable trip to Seattle's Boeing Field. The flight with Concorde service was supersonic across the Atlantic to New York, then, due to U.S. regulations, it was subsonic to Seattle, which allowed for more drinks and good food.

The arrival was beyond expectations. Tens of thousands of spectators watched as the beautiful bird approached, landed, and disgorged

Museum Wins State Centennial Award

The Museum of Flight was honored to receive one of the five major Washington State Centennial awards presented by state leaders in November. These five top awards were chosen from among over 5,000 individuals, companies, and organizations involved in the Washington State Centennial. The Centennial award, given to recognize the museum's leading role in developing *Wings Over Washington*, a tribute to our state's flight heritage, provides a fitting recognition for the many volunteers and staff members who devoted months toward this statewide project, and to Wings Over Washington's corporate supporters. Wings Over Washington, which on July 29, 1989 involved an es-

entire state. The governor also delegated chief responsibility for Centennial planning to his wife, Jean, and Ralph Munro. Shortly afterward, the Museum of Flight began to play its key role in developing and promoting special flight-related Centennial events throughout the state.

The other four Centennial awards were presented to: Frank B. Russell Company (for sponsoring the awards ceremony), The Boeing Company (for sponsoring Chautauqua and supporting Wings Over Washington), Burlington Northern (for underwriting and restoring the Centennial Games Train), and Seattle First National Bank (for sponsoring "Over Washington" the centennial's official film and book).

its manifest of contented passengers with Washington First Lady Jean Gardner in the lead. The media was energized, the crowds thrilled, and at this moment it was clear that the Museum of Flight was the right kind of institution in the right place.

The Flight of the Century attracted some of the largest crowds in history to Boeing Field and garnered media coverage, both print and broadcast, across the country. During the Wings Over Washington events, the museum, through its new foundation structure, funded, promoted, and organized another dozen or more flights of the century around the state. This coordination made the celebration meaningful, resulting in a great tribute to the anniversary. This event not only paid for itself but generated net revenue. That was a relief to those who took a mighty risk.

On November 10, 1989, Lovering was invited to a Centennial Award reception at the Governor's Mansion and was surprised to be the recipient of one of the top awards on behalf of the Museum of Flight. Of the 5,000 Centennial events, Wings Over Washington with the Flight of the Century was by far the largest, and it was the recipient of one of only five major awards, the only nonprofit to be so recognized.

In presenting the prestigious Centennial Award, Secretary of State Ralph Munro pointed out that this flagship event reached an estimated 10% of the statewide population with the dozens of related celebration flights held around the state. The museum's trophy room now proudly displayed the Washington State award for the best nonprofit event of the Centennial. In these early years the museum had organized its largest attendance at the B-17 50th anniversary, and now what was proclaimed the most-attended event in state history. This is the kind of performance that a museum might hope to accomplish once in a century.

The museum is recognized for the Wings Over Washington events in celebration of the State Centennial. From left, Governor Booth Gardner, Putnam Barber, Jean Gardner, Howard Lovering, and Secretary of State Ralph Munro.

Howard Lovering
Executive Director, 1977 through 1991

In 1975, I was introduced to PNAHF as a Boeing Company loaned executive. Reporting to then company treasurer Jack Pierce, I was to conduct an assessment of the community's need for a flight museum. With the task complete, I returned to Boeing in 1976, while serving on the PNAHF board and continuing to volunteer with planning. William M. Allen, Boeing company chair, offered me the position of museum Executive Director in February 1977, and I accepted. What was expected to be a leave of absence for a year lasted 15, and I never returned to the company.

The experience was an adventure, but always with an exit plan. Most of us on the team considered hitting the silk more than once. Somehow, we managed to push ahead and accomplish some unlikely tasks to locate the museum, save the Red Barn, devise a concept, design the complex, and start the fundraising. The journey was a broadside experience for me and I suspect would have been for anyone. Along the way, I kept a diary of best practices and some ideas for managing a museum. Surviving the building program found me determined to take a shot at running the place.

Operating a museum is neither a sprint nor a distance race. In sports parlance, it is more akin to the decathlon. The key is to commit the best possible performance in all categories. That is how a museum can excel, and that is what has transpired at the Museum of Flight.

Coming off the fun and proud accomplishment of opening each of the phases, with governors and community leaders, vice presidents, and astronauts participating, it is a thud of reality to welcome the public. Suddenly, the party is over, and there is just a lot of hard work and cleanup until the transition is made to enjoy the routine. As we found, it takes a team—board, staff, volunteers, and supporters—with audience feedback to get bearings. The museum was fortunate to have energetic and loyal people to make that transition.

Executive Director Howard Lovering at Claire Egtvedt's desk in the Red Barn.

Our collective objective was to prove we deserved this promising institution by doing, not talking. Sure, we all wanted excellence, but we had to start by being the best we could be. We had many deficiencies, as would any new institution. Accomplishments were in degrees, leveraging what we had to get what was needed.

During initial operations, strategic plans were prepared to enhance performance. The team identified artifacts considered essential to the themes and began search-and-rescue efforts. Restoration of inventory was ramped up, a long-range task. Budgets were prepared to enhance exhibits and became part of a follow-on campaign. Public programs were tested to find those that were most instructive and attractive. All of this constituted a strategy of careful phasing that has flourished at the museum over the years.

A smart example of this planning was the attention conferred upon space themes. A question was whether or not the museum could ever present a suitable story on the subject, and obtain the necessary authentic artifacts from a scarce and dwindling supply. Because the foundation stressed education as paramount, it was crucial to involve space flight to attract young people. Initially, these stories incorporated just a few artifacts, including replicas. The participation of the Mercury crew at the grand opening highlighted the commitment. This welcoming of the space community continues at the MOF and facilitates the attraction of support and valuable artifacts.

Finally, the museum had to prepare for long-term sustainability simply by giving no concessions. From the beginning, the foundation adopted a statement of purpose that was bold. Over the years the mission has been amended in line with changing circumstances, each time defining a course to excellence. The Museum of Flight has fulfilled expansive expectations.

Personally, I think back to the building and operating with fond memories of the people I worked with daily. Several of these colleagues had museum experience and were mentors, but most of us were inexperienced. What we lacked in museum savvy, we made up for with energy and imagination and a team spirit that I have not encountered elsewhere. We thought we could do almost anything expected of us, and we tried and often succeeded.

Most of these friends have gone on to prestigious positions, including heading up other museums, founding businesses, running a school system, nonprofit development, consulting, and leadership in national organizations. We stay in touch and still get together on occasion to share stories of our early adventures at the Museum of Flight. I have noticed that the stories get better as we age. These continuing relationships give me the best of all memories.

— Howard Lovering

Cockpit of the
museum's Boeing B-17F.

CHAPTER THIRTEEN

A New Flight Plan

Although the museum's artifacts aptly interpreted the themes of commercial, military, and general aviation, its space component was noticeably weak. This deficiency needed some first aid while the museum searched for artifacts.

During one of his trips to Washington, D.C., Lovering visited the headquarters of the Challenger Center for Space Science Education to discuss the possibility of hosting a Challenger Learning Center (CLC) at the Museum of Flight. The CLCs were a new and growing network of space exploration education. They featured simulated missions and team-building of particular interest to students. Although based in reality, these centers did not require artifacts and for the museum represented an opportunity to upgrade its space themes instantly.

Lovering learned that Seattle's Pacific Science Center (PSC) was the favored target for the region. Upon his return, he talked to George Moynihan, CEO at PSC, hoping to effect an agreement of some sort. Moynihan had a valued space education program, but neither museum had the money to establish a CLC, a $1 million investment. Moynihan agreed that due to this impasse, Lovering could continue to promote the location at the Museum of Flight. With that agreement between museum directors, Lovering asked for and received board support to make a pitch for the Challenger Learning Center.

In early 1991, the Museum of Flight was selected for the first CLC in the region. This coup was a significant boost to the space themes the museum needed to expand. Lovering thought the Challenger program was formative, clunky in execution, but that it was expansive in its network of facilities. It was the kind of program that the museum could not produce and therefore was a value at the price. Although the museum did not have the money, the deal was made to implement the CLC and start the funding process.

Focusing on strengthening other weaknesses at the museum, Lovering landed on the idea of securing and expanding the restoration function at Paine Field in Everett, Washington. The popularity of the Boeing factory tours, the annual Paine Field Air Show, and an airport with a large land base serving the growing population to the north drove the notion of developing the restoration center into an MOF satellite attraction. After some review, in 1990 he proposed an effort to broaden the service into the National Flight Interpretive Center (NFIC), to include not only an enlarged restoration hangar with public access but also Boeing plant tours and flight interpretation, something of national significance.

By 1991, NFIC was in the works. The museum was renting the Paine Field facility, and the owner offered it for sale at a reasonable price. Absent cash, the foundation looked for favorable financing, and the Boeing 747 provided an answer.

Boeing had a renewed use for its prototype 747 that it had gifted to the foundation in 1989. The museum was now poised to lease the aircraft back to the company for new generation engine testing. It was an opportunity to produce income while also validating the purpose of this historic test bed aircraft. This arrangement offered collateral for a favorable bank loan to purchase the restoration facility. In turn, subletting a portion of the hangar to Boeing for offices reduced the museum's monthly expenses, resulting in a new asset with a cash-flow improvement. This purchase created a much better restoration space to accommodate visitors and build an attraction at Paine Field. The planning committee and the board reviewed and authorized the initiative.

The complex was designed to support the factory tours, add facilities for the museum's larger aircraft, and enhance the visitor experience at the restoration facility. The museum pitched the combination of historical and contemporary flight themes to the National Park Service, which was seeking a center to celebrate aviation. The regional office responded favorably, as did various members of the state and federal delegations in promotions to government partners.

In August 1991, hundreds of museum supporters gathered at the Restoration Center at Paine Field to hear the plans for the National Flight Interpretive Center. Joined by U.S. Senator Brock Adams, Snohomish County Council members, and the airport manager, MOF planning committee chair Don Van Blaricom announced, "This partnership between the Museum of Flight, Snohomish County, and the National Park Service will result in a very popular visitor attraction and cultural resource that is of national importance." The idea gained momentum during 1991, then would take on several additional iterations over the years.

The Tuskegee Airmen continued to return to the museum over the years and have always been a crowd favorite.

Historic aircraft and aviation pioneers found their way to the museum. Among the distinguished speakers was retired General Benjamin O. Davis Jr. For two days in September 1991, he spoke to public audiences and private groups about his experiences as the commander of the famous Tuskegee Airmen of WWII. The museum and the local chapter of the Airmen sponsored the event, and General Davis, the first African-American Air Force general, proved a dynamic personality. This initial visit cemented relations with the Tuskegee association, which would years later formally partner with the museum.

For many years, the museum actively sought a loan of one of the Lockheed Blackbird family of aircraft, not knowing when one might be available. The effort was rewarded in 1990, when the small fleet was declared surplus to U.S. Air Force needs.

The Blackbird aircraft went on to their new lives as star attractions in aviation museums, with a rare M-21, one of two drone-carriers manufactured—and the only survivor—offered to the Museum of Flight.

The MOF spent months organizing the transfer of the Blackbird from Mojave, California, to Seattle. The museum curator of restoration supervised the dismantling and move, with support from staff and volunteers. All of the equipment was donated. Five Kenworth trucks, three of them rated for handling oversized loads, hauled the aircraft parts up the I-5 corridor. The journey north took 12 days, producing what became a long-distance road show as crowds gathered in cities and rest stops along the way. Hundreds of enthusiastic spectators greeted the fleet and crew when they pulled into the museum parking lot on September 17, 1991.

The Blackbird is a spectacular performer that looks as dramatic as it flies. It is one of the world's great aircraft, the highest-flying in the museum collection and the fastest air-breathing aircraft in flight history. The Blackbird is a plane that defines a gallery, one of those particular artifacts that can boost attendance and intrigue both young and old. It is an aircraft around which storytelling is endless. Its D-21 Drone was added in 1994 to make the exhibit even more authentic.

Top: Arrival of the Blackbird from Mojave, California, 1991.

Bottom: Curators Victor Seely and Jay Spenser imitate the museum's logo For Future Generations.

In 1991, a MiG-21 was purchased through the work of Jim Blue, and the U.S. Air Force loaned the museum an F-4 Phantom. The museum also took delivery of its promised B-52G on September 23, 1991, when it flew into Paine Field. Former Boeing executives Maynard Pennell, T. Wilson, George Schairer, and Guy Townsend, who had worked on this aircraft early in its development, welcomed it at the acceptance ceremony. The B-52, still in military service, speaks to the high quality of Boeing design and engineering in its incredibly long life.

In the latter part of 1991, the new Apollo exhibit was well along in research and design. The first of the museum's space displays was made possible with the 1986 acquisition of the Apollo Command Module. The museum had fortuitously located this moldering yet precious artifact in a vacant lot in Fort Worth, Texas. The module then underwent several years of careful restoration with craftsmen at International SpaceWorks in Hutchinson, Kansas.

Lockheed M-21 Blackbird

World events moved quickly in the Cold War. Fortunately, Lockheed Aircraft responded just as quickly to create a 1960s flying machine that's still the fastest air-breathing airplane ever produced. Different versions have different designations, but all are known as Blackbird.

In the 1950s, Lockheed had introduced the F-104 Starfighter, a Mach 2 jet fighter. That same decade, the company began deliveries of the classified U-2, a jet-powered glider designed for long-range photo reconnaissance from extreme altitude. The U-2 combined a modified F-104 fuselage with very long wings, allowing it to cruise at 70,000 feet (21,336 m).

The U-2 was a creation of Lockheed's famous Skunk Works. Officially known as the Advanced Development Projects division, the Skunk Works was a miniature company-within-a-company that gathered together Lockheed's best and brightest to conceive, test, prototype, and fly highly classified aircraft of advanced capability. At its helm was legendary designer Clarence "Kelly" Johnson.

On May 1, 1960, a Soviet missile downed a U-2, generating international headlines and calling a halt to U.S. overflights. The United States found itself suddenly in need of a stealthier, more capable, and more survivable aerial reconnaissance plane.

The U-2 was slow even by subsonic jet standards, and its radar cross-section was too large. Fortunately, the Skunk Works was already working on the solution. What coalesced on their drawing boards looked like something off the cover of a sci-fi magazine. The design featured a flattened fuselage projecting far forward from a delta wing whose integral engine nacelles were topped by inboard-canted vertical fins.

Many hurdles had to be surmounted to create the single-seat Lockheed A-12, as this new spy plane was designated. One was learning how to build almost the entire airframe out of titanium, a notoriously difficult metal previously used only in specific locations where great strength, light weight, and resistance to extreme temperatures were required.

The A-12 flew in April 1962 at Groom Lake, the secure Nevada desert test facility also known as Area 51. By early the next year, A-12s were flying at more than Mach 3 and climbing above 85,000 feet (25,908 m). Of 18 built, 13 were delivered to the U.S. Central Intelligence Agency through 1964. As was the case with the U-2, the CIA operated its A-12s using civilian pilots with prior military flight experience.

Although the A-12 flew faster than a rifle bullet at the very edge of space, it was not ultimately used for its intended role. Soviet radar detection and countermeasures had become ever more formidable even as emerging U.S. satellite reconnaissance capabilities undercut the need for airspace penetrations.

Three A-12s, procured with USAF funding, were modified to YF-12A configuration and evaluated in the interceptor role. This experience prompted the Air Force to order 32 additional Blackbirds under the designation SR-71. Last flown by NASA, the last SR-71—the ultimate Blackbird version—retired in 1999.

Back in 1963, two A-12s were modified on the assembly line to become perhaps the most exotic aerial reconnaissance planes in history. Designated the Lockheed M-21 Blackbird, they featured a pylon on their backs for mounting the even faster, even higher-flying Lockheed D-21 drone. The two-seat M-21's second cockpit accommodated the launch control officer. One M-21 was lost to a crash, leaving the Museum of Flight's beautiful example as the last of its kind.

The M-21 with its D-21 drone is a magnificent backdrop for special events.

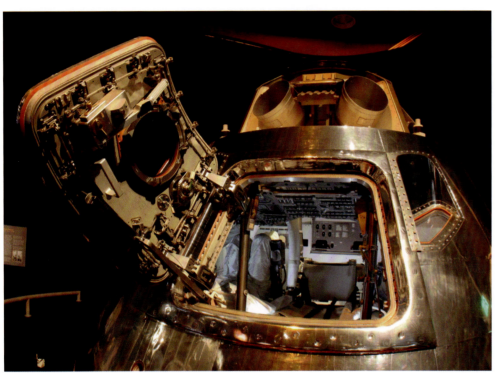

Top: Arrival of the Apollo Command Module in 1986.

Bottom: The museum's Apollo Command Module 007A with its complex door lock/ release mechanism.

With donations and loans of additional small artifacts, the Apollo story was expanded into a large space-themed exhibit, and in January 1992, *Apollo: The Journey Begins* opened to enthusiastic crowds. More than 180 objects, photos, and models augmented the refurbished Apollo Command Module. With the help of design and production consultants, the staff prepared one of the museum's largest exhibits to that time. The rich layering of interpretation brought out themes and stories that appealed to a broad range of interests and cemented the space theme. The exhibitry exceeded what the print promotions had promised, that it highlighted "the cultural and environmental impact of manned spaceflight and the political, social, and artistic changes it wrought."

By the time Lovering left the museum in December 1991, the organization was humming along in operations, working on new projects and acquiring valuable artifacts. Lovering had served over two decades, from early planning in the 1970s through the major capital campaign and building of the museum, and into thriving operations.

With the change to a new regime, it was important to take stock of what the museum had learned and accomplished, and what was yet desired to take it to its full potential. This transition was opportune for the museum, occurring during a period of rest from capital fundraising, and with several years of operations under its belt. However, debt remained for completion of the building program, and already committed improvements, including the Challenger Learning Center, were not yet fully funded. The new CEO would need to take a serious inventory.

Ralph Bufano began work as Lovering completed his tour as director in December 1991, taking over in January 1992 with a new title, President and CEO, more in keeping with a business model for the museum. Bufano had a solid museum background that included the position of executive director of the Experimental Aircraft Association (EAA) Museum in Wisconsin.

The EAA is one of the largest aviation organizations in the world, with 180,000 members across the country. EAA featured an annual premier fly-in and air show, which provided a host of valuable contacts that Bufano could use over the years of his tenure at the MOF.

While working as director of the Ward Museum of Wildfowl Art in Maryland, Bufano had seen the ad for the MOF director position and called to ask Lovering if he should apply. Lovering encouraged Bufano and reminded him that only a day or two remained before the application period closed. Bufano submitted a hasty proposal and soon received word that he was a finalist. When selected, he was asked to get to work as quickly as possible.

Bufano hit the museum floor running. He was aware of the big task ahead just to continue the excellent programming and public education, and T. Wilson and Bill Boeing Jr. briefed him on the expectations of leadership in the community.

Bufano suggested that the MOF, as good as it was, could be better, so much better that he thought there was a chance to be the best flight-education institution in the world. This comment received attention and began to be discussed informally, then formally. Several board members asked if it was true that he thought the MOF could attain this high station, and he answered, "Yes, I do, and all it will take is staff and money." This frank assessment did not deter the board or slow the conversation.

The museum had already organized and committed to the Flight and Operations Campaign, a list of improvements that served as the prescription for follow-on capital funding requirements. Over the longer range, the museum required an upgrade in collection storage and maintenance. With large artifacts at risk in the weather, an additional physical plant, located somewhere, was needed for their protection. In addition, the education mission required enhanced interpretive galleries for aerospace history and technology. It would take imagination and perhaps a miracle to realize these improvements on the limited land base.

Significant at any time, these tasks were even more daunting in a time of several economic crises. Recessions reduced attendance, which lowered earned income, and limited the capital resources from the donor community. Budget reductions resulted in employee layoffs and decreased compensation at a time when there was so much work to be done. Nothing was easy during this period, which was in effect more of the same in an aerospace economy that always had ups and downs.

The museum and its growing cadre of supporters in the community rallied, and with a few breaks along the way, turned this period into one of major growth in quality and quantity. With guidelines in hand, the board worked with the new CEO to build up the capability to fulfill the improvements, expand the board, and create a committee structure tasked with assignments.

The museum set out to take the institution to a new level of collecting, interpretation, education, and research.

Debts were paid on time, visitor amenities improved, the collection expanded, and displays enhanced. Capabilities were added in education, conservation, exhibit design, and fundraising. The enlarged board recruited talent to a remodeled committee structure in direct support of planning needs. The active board and committee organization might be one of the most significant of all the changes made during this period, and continues to serve the museum well. None of this came easy, but the story of starts, stops, and setbacks ended with ultimate accomplishment. As a result of this period of transition, the museum was poised for its next stage of development.

Bufano was delighted that the museum was designated as the location for the first Challenger Learning Center in the Northwest. However, a sizable payment was due to the Challenger Foundation, and there was no money in the budget to meet the obligation. Bufano organized a funding campaign and traveled to Washington, D.C., to buy some time for the initial payment. He negotiated a new schedule with Doug King,

Top: The first class of Red Barn docents, gathered in 1993 on the 10th anniversary of the Red Barn opening. Jim McDonnell (top row, second from right) was still a docent in 2016.

Bottom: Newly installed Challenger Learning Center, 1992.

Challenger CEO, who would become the MOF CEO nearly 20 years later. With this relief, the museum completed the funding for installation of Challenger, an improvement that was to leverage a higher level of service for the education programs.

Challenger Learning Center was installed in the Great Gallery in 1992 and was an immediate success. Space flight was fun and instructive to the young as well as for adult group activities. The simulated experiences were spiced up with real-time communications to astronauts and space conferences. This new facility bolstered the space themes that were beginning to flourish.

The international network of Challenger Learning Centers provided a group of partners with whom to share programs, and the museum education staff plunged into enhancing the experience and communicating with others in the network. The MOF education department and volunteers took on a leadership role in this association, developing one of the finest and most sophisticated Challenger units.

The funding for CLC was accomplished with one of the mini-campaigns that would prove successful over the next decade, each targeted to a particular need and packaged for specific donors. In 1993, the associated Murdock Theater, a multi-use facility, opened adjacent to Challenger Learning Center, expanding the programming opportunities. The museum was booked months ahead with these space travelers. That year the State of Washington appropriated a challenge grant of $800,000 for an expanded exhibits program, one of the early capital contributions from government.

Immediately at hand was the obligation to build out the food-service facility that would complete the Great Gallery complex. The museum was obligated to build the shell structure, catering kitchen, café, and banquet room with support spaces.

Students participating in a CLC mission simulation in the early 1990s.

Wings Café at the Museum of Flight.

King County and the airport had attracted few responses in previous advertising for an aviation-themed restaurant. When the venue was transferred to the museum, the request for food-service proposals received promising responses from skilled and experienced restaurateurs. The Museum of Flight Authority evaluated proposals, providing oversight to both county and museum interests in the land and improvements and associated details.

The process resulted in an operator award to McCormick & Schmick's restaurants in 1991. Their proposal provided the best terms for a quality venture in which the operator would invest some of the capital for improvements, pay a fee for space, and run the business to museum guidelines. This favorable agreement resulted in one of the first full-service café and catering functions within any Northwest museum. This lease would be reevaluated and renewed over the years to the current day.

The board judged this to be an improvement not appropriately built with donated funds. The financing package ultimately incorporated short-term loans, a significant amount provided from the airport reserve funds that were scheduled for repayment with earnings from food service and events. With construction financing for the restaurant addition in hand, design began with Gregov Architects. Lidija and Ivo Gregov had served with Ibsen Nelsen and Associates as project architects for the Red Barn and Great Gallery. Construction started in 1993 and was completed in 1994, adding 17,430 square feet of support space and enhancing the museum's capabilities for functions large and small. The café offered snacks and meals for the visitor at what is a destination location in an industrial area. The catering function handled food-service requirements from simple box lunches for small group meetings to large and lavish banquets. With these facilities, meeting and event businesses were cooking.

The final component planned for the Great Gallery was the control tower exhibit. In all of the various concepts for the museum, the planning team had always suggested a mock control tower, and this was the time and opportunity. In 1995, with an initial contribution of $170,000 from King County, this addition was launched. Again, Bill Boeing Jr. stepped up, with an $800,000 gift to fund the balance. Boeing Jr. had always thought a real tower experience would be useful for education. He had a particular interest in air-traffic control from the work of his uncle Thorp Hiscock, an aviation pioneer who developed the first air-to-ground radio and early autopilot. With equipment donated by the FAA and additional contributed services, the $1.5 million project was soon under construction. It opened to fanfare in March 1997.

Lidija Gregov, working once again with structural engineer Jack Christiansen, designed the tower to fit into the airfield side of the Great Gallery. The observation area was large enough inside for crowds and displays. Proportionate with the Great Gallery, separate but connected, it featured similar materials in a noticeable round form integrated into the long rectangular space overlooking the runways.

The 30-foot tower explains ground and air control at the active airfield and the theme of highways in the sky. The view is dramatic, and the exhibits interpret air communications, from flags and bonfires to radar and satellite technology. This reality-based experience is regularly updated and has proven popular with adults as well as children.

The American Institute of Architects (AIA) selected the tower as Project of the Month in October 1997. The jury commented, "There is a dramatic feel without a large scale . . . a nice add-on to the large, distinct structure, and good proportions. It gives a sense of flight, historic to modern."

Another of Bufano's goals was to augment the staff with more professional museum practitioners. He was pleased to recruit Jack Hilliard from the National Museum of the U.S. Air Force as a new curator. Upon Hilliard's retirement in 1995, Bufano hired Dennis Parks, another seasoned curator, who continued to build the library and archives.

Bufano also decided it would be cost-effective to make exhibit design and development an in-house function. He recruited exhibit designer Chris Mailander, who developed a top-notch design and production department.

Given the demand for restoration and maintenance of exhibits as the collection grew rapidly, Bufano found experienced supervisors to run the Paine Field Annex. Tom Cathcart took over the function in 1995 and built a professional cadre to organize and address the growing number of restorations.

Left: Visitors track air traffic on Boeing Field from the control tower.

Right: The control tower, added in 1997, was artfully integrated into the Great Gallery structure.

Bottom: Jack Hilliard, Museum of Flight Curator, 1993-1995.

With these professional upgrades and the Challenger Learning Center, the education staff was continually expanded, bringing in more teaching and curriculum experience.

In 1997, the board adopted a new vision statement: *To be, by the beginning of the 21st century, the foremost independent educational air and space museum in the world.* As part of the "Vision 2020 Statement," the associated performance goals and objectives shaped the plans to take the museum into the new century.

In September 1999, The Boeing Company donated a six-acre parcel to the museum. This land was located across the arterial to the north and west of the main campus. With this new property, the executive committee requested a revised museum master plan and development process. The staff and consultants worked with the planning committee on this assignment to reevaluate the mission of the institution in reference to its current operations. This review inspired changes that in turn led to the expansion of both staff and the physical facilities.

This mission enhancement resulted in a requirement for an enlarged "education center" to serve informal education, curriculum design, teacher training, and lifelong learning through flight themes. The idea at the time was to facilitate this educational expansion at both the Boeing Field campus and in the market to the north, at the emerging complex at Paine Field.

President and CEO Bufano with a model of the 2002 master plan.

The museum contracted with a national market research firm to conduct an audience assessment. Many of their recommendations were improvements that had been long discussed but never implemented. Now was the time to make the changes.

Although the MOF was recognized as a top-tier museum attraction, the assessment identified a need for new ways to grow the audience, especially in regard to residents. The museum was perceived as narrow in its themes, more comfortable for aviation buffs. The proposal was to add more interpretation and interactive experiences to broaden the audience. The pitch was to reach out more to the young with space exploration, to females, and to families with attractive exhibits that were engaging and compelling. The surveyors also suggested that the existing educational and entertainment features would need enhancement and promotion, requiring a much larger budget.

The assessment also addressed the commercial transport component and the nagging concern about protecting and exhibiting the larger aircraft, some of which were among the most valuable of all museum artifacts. Over time, the need for a large facility of this kind was discussed in several forms, ultimately resulting in the Aviation Pavilion, completed in 2016 on the west campus of the museum complex. On this long list of to-do items, there was not one that was just for show or fancy. All were practical needs to take the museum to its next level of service.

The summary of the master plan report suggested that the complex would be most economical if built in total, including the bridging of the busy public arterial road to reach facilities on the west campus. Optimistically, its completion was suggested for the 100th anniversary of flight, in December 2003. Upon review, the cost of such an expansion was judged unfeasible, and the schedule unrealistic. The planning team went back to the drawing board to find the optimum next step that was practical for funding.

The museum's collection grew significantly in quantity and quality over the next decade of expansion. Several additions were in the pipeline and just awaiting restoration. But there were also new pieces of tremendous value. Some of these were targets of the newly enlarged collection committee; others came along as surprises.

On occasion, there would be little time to organize an acquisition procedure and budget because the market for these rare artifacts was fluid and changing. Opportunities were weighed for priority and then a strategy was set for funding the item on a case-by-case basis.

Top: Volunteers refurbished the 1971 Boeing Lunar Rover mockup as an addition to the museum's space themes.

Bottom: The Molt Taylor car on loan since the opening of the Great Gallery was purchased and gifted from the Friendship Foundation on the 10th anniversary of the record-breaking flight.

A collecting flight museum is often judged by the value of its artifacts, with the worth of each determined by its rarity and role in the advancement of aviation. It is paramount to get first-off models, even a prototype. An artifact that carries important operational stories is even more valuable and can become an item of competition. The Museum of Flight is fortunate to have numerous aircraft that are the envy of many institutions.

The Goodyear FG-1D, so beautifully restored over the early years, became a highlight display during this period. The Molt Taylor Aerocar, originally on loan, was purchased with a donation from the Friendship Foundation on the 10th anniversary of the historic 747 Friendship One around-the-world flight. The de Havilland Comet 4C came to rest at the museum, a milestone commercial aircraft donated by Everett Community College.

A McDonnell F-4C Phantom II, Boeing Lunar Rover, Grumman A-6E Intruder, and Boeing VC-137B also joined the collection. It seemed at this point that the collection of artifacts was outstripping the facility capability. It was fulfilling the prediction for the new museum and much more.

Boeing 707 Air Force One

The Boeing 707—history's first truly successful jetliner—entered service with Pan American World Airways on October 26, 1958. Debuting across the North Atlantic, it wowed the public and became Boeing's first-ever smash hit on the commercial front.

Later that same year, the U.S. Air Force acquired three specially built 707-120s to serve with the U.S. Presidential Flight. Delivered in 1959, these special-air-mission VC-137As were designated SAM 970, SAM 971, and SAM 972. Only when the president was aboard would they use the call sign Air Force One. The press and public, however, were quick to indiscriminately apply this term to all of them.

The VC-137As whisked presidential transport into the jet age. With rakish lines and swept wings that bespoke world-leading technology, they further heightened the glamour and prestige of official state visits by the chief executive of the United States.

The Museum of Flight proudly displays SAM 970, the first presidential jet, which is on long-term loan from the National Museum of the U.S. Air Force in Dayton, Ohio. Like her sisters, SAM 970 is fitted with what was then state-of-the-art communication equipment, allowing her to serve as an aerial Oval Office.

During her years in harness, SAM 970 carried presidents Eisenhower, Kennedy, Johnson, and Nixon as well as countless VIPs, one being Nikita Khrushchev. On a visit to the Museum of Flight, former U.S. Secretary of State Henry Kissinger reminisced about his heavy reliance in particular on SAM 970.

Theodore Roosevelt was the first U.S. president to fly, although by the time he took his hop in a Wright Flyer he was out of office. During World War II, his distant cousin Franklin Roosevelt spanned the globe by air. Pan Am's *Dixie Clipper*, a Boeing 314 flying boat, took him to Morocco in January 1943, and his personal presidential plane, a Douglas C-54 Skymaster called the *Sacred Cow*, carried him all the way to the Crimean Peninsula and back in February 1945.

President Harry Truman had a Douglas C-118 Liftmaster, the military version of the DC-6. Named *The Independence* after his hometown in Missouri, it provided comfort lacking in the unpressurized C-54, a military version of the DC-4. Technology continued to improve, and President Eisenhower enjoyed speedier travel aboard two Lockheed C-121 Super Constellations known as *Columbine II* and *Columbine III*. Then came the Boeing jets, and the difference was like night and day.

The Boeing VC-137As arrived during Eisenhower's last full year in office. Noisy as those early jets were from the outside, in the air it was another story. Gone were the fatiguing noise and vibrations of piston-powered propeller airliners. Better still, the speedy Boeing cut flight times in half. Before leaving office, Eisenhower made a goodwill tour of 11 Asian nations, covering 22,000 miles (35,000 km) in just 19 days.

In 1962, two VC-137Cs were added to the Presidential Flight. Larger and featuring greater range, SAM 26000 and SAM 27000 took over first-line duties in the administration of John F. Kennedy. These newer ships—military versions of the 707-320 Intercontinental—also introduced a gorgeous new blue-and-white paint scheme that was retroactively applied to the earlier Boeing model and is still used today on the Boeing 747s that transport the U.S. president and other senior officials.

SAM 970 VC-137A is delivered to the Museum of Flight, 1996.

In 1996, there was the triumph of acquiring SAM 970 "Air Force One," or the Flying Oval Office, made available on loan from the collection of the National Museum of the U.S. Air Force. This Boeing VC-137A was the first jet aircraft in presidential service, used by Eisenhower, Kennedy, Johnson, and Nixon. The aircraft carried a variety of important world leaders, making it a visitor favorite today.

The collection gained an Antonov An-2 aircraft and a rare Douglas DC-2 in 1999. Board members Bruce McCaw, Joe Clark, and Clay Lacy discovered the DC-2 in California and acquired

it for the museum. Lacy took on this important project and supervised its restoration to flight status. The Douglas DC-2 represented a major coup for the museum, adding, at the time, the only airworthy version of this historic craft in existence.

The museum's rich storytelling was ramped up by the magnificent restoration of the B-17F that was completed and delivered in flyable condition during this period. "Swage" Richardson would have been proud to see his incredible "Boeing Bee."

The B-17 restoration is a testament to the talents and contributions of many volunteers and the in-kind services of The Boeing Company. Pat Coluccio and Tom Elliott of The Boeing Company took on the challenge and welcomed 100 committed volunteers who spent more than 100,000 hours on the task. The crew found space at the Boeing plant in Renton and went to work. The initial inspection revealed that the condition was

worse than expected, with damage to the structure, loose panels, and massive corrosion from its years carrying water and chemicals. Controls were missing, and most armaments were stripped. The engines had to be overhauled or replaced. The team knew this was going to be a long haul, and it was.

Materials were inventoried, hundreds of parts were found or manufactured to specifications, the armaments were faithfully replicated, and every inch of the aircraft was inspected and restored. Over the seven years of this intense effort, progress hummed, and when the B-17F landed at Boeing Field in May 1998, it proudly displayed the *Boeing Bee* insignia as a badge of honor. As Pat Coluccio explained, the value of the work is inestimable, going well beyond millions of dollars to an asset that is timeless and priceless. The B-17F is another artifact that defines the museum and attracts continual attention.

Then there was the capture of the original Boeing 737, which took to the skies on its first flight on April 9, 1967, with Brien Wygle and Lew Wallick in the cockpit. After its successful performance in certification, the prototype 737 was taken into NASA inventory in 1974 at Langley Research Center as NASA 515, where it served as a flying lab to test aircraft systems and traffic control. The 737 did a lot of work on identifying and avoiding wind shear and contributed to flight safety. In 1997, the museum received the aircraft from NASA on permanent loan, and it was flown to a Boeing facility in Moses Lake, Washington, for restoration. NASA 515 made its final flight into Boeing Field in September 2003.

This aircraft, representing another first off the line, joined the 727 and 747 prototypes, which were awaiting necessary restoration work to prepare them for their final display at the museum. The trio of test beds is significant to what has become one of the most valuable and extraordinary commercial aircraft collections in the world.

Trustee and Boeing test pilot Brien Wygle at the 747 donation ceremony, March 28, 1990.

This remarkable era of acquisition and restoration was not just lucky but earned with the efforts of the board, staff, and supporters. Brien Wygle, longtime chair of the Collections Committee, and other members had both local and global connections and kept up a constant surveillance for what was available. This group of volunteers typified what was best in this robust committee-driven organization. Their efforts continued the expansion of the museum's collection beyond expectations.

The two main antagonists of the Battle of Britain—the German Messerschmitt (prop on left) and the British Spitfire—face off in the entrance of the World War II gallery of the Museum of Flight's Personal Courage Wing.

CHAPTER FOURTEEN

A Flock of Warbirds

Moving in artifacts, opening a new exhibit, hosting a traveling exhibit, or producing a special event are all opportunities to promote the museum as a happening place, and these were regular occasions.

The museum celebrated the 50th anniversary of the B-29 in 1992, attracting 8,000 veterans and national media attention. This event showcased the visit of *Fifi*, one of the last flying models of this historic World War II bomber. Along with the crew members and their families, Boeing engineers shared stories of the intense effort to design and produce the workhorse. The B-29 was the first pressurized bomber, flying twice as fast and far as its predecessor B-17. In service, it developed a reputation for performance and durability, often returning from a mission severely damaged. In use until 1960, the bomber made a difference in the long-range WWII missions in the Pacific Theater.

In the summer of 1992, the U.S. Navy Blue Angels began using the museum as their base when the F/A-18s and stunt crews made their annual appearance at Seattle's Seafair. This marine festival was established in 1950 and had swelled into three weeks of fun events and parades, culminating in early August with the hydroplane races on Lake Washington. Since 1972, the high-speed maneuvers of the Navy and Marine pilots had thrilled the spectators lining the lake, and now the public could see them up-close on the Boeing ramp just south of the museum, a partnership providing education, entertainment, and drama.

The Resurs 500 capsule's debut at the Museum of Flight also made national news. In mid-November 1992, a Russian rocket launched the Resurs 500 into polar orbit from the secretive Plesetsk Cosmodrome in Northwestern Russia. The 5,000-pound sphere made 10 circles of Earth before splashing down in the Pacific Ocean on November 22, approximately 100 miles off the coast of Grays Harbor in Washington State. Helicopters flew local dignitaries and journalists to the 680-foot tracking ship *Marshal Krylov* to witness the hardware's retrieval in gusty winds and rough waters. This feat marked the first time Russia recovered one of its satellites at sea, as well as allowing foreigners aboard one of their naval ships.

Top: 50th-anniversary celebration for the Boeing B-29 Superfortress, 1992.

Bottom: Resurs 500 space capsule, recovered off the coast of Washington State, just before its official donation to the museum.

Traveling through the Strait of Juan de Fuca and south through Puget Sound, the *Krylov* docked at Pier 42 for a public open house. After starring in Seattle's holiday parade, the Resurs, orange and black from its fiery re-entry, traveled atop a Russian amphibious vehicle to the Museum of Flight, where its cargo of gifts, a toy stuffed dog named Digswell, and its messages of peace were unloaded. Named the "Europe-America Spaceflight 500," the event coincided with the 500th anniversary of Columbus' voyage to the New World.

Working with Russia's Foundation for Social Innovation, local promoters Bob Walsh and Associates organized the event with the objective of cultivating mutual business relationships as an extension of Seattle's 1990 Goodwill Games. Walsh and trustee Bruce McCaw appreciated MOF's service as a venue for the games, which made it the perfect recipient for the Resurs capsule. This artifact assisted in enhancing the space exhibits and gaining international partnerships.

The restored 247D flew from Paine Field into Boeing Field as a feature attraction of the Emerald City Flight Fest in 1994, proving continuing interest in the museum's collection and value as a setting for the aerospace and commercial transportation industry.

The 1994 Seaplane Fly-In added Renton Airport to enable water landings and takeoffs, greatly enriching the event. That year, Joann Osterud donated her Stephens Akro aircraft she used for years in aerobatic competitions and as a popular airshow performer. A museum exhibit with the aerobatic plane was installed at the Washington State Convention Center on temporary display in 1995, again building brand and recognition. In 1995, the museum was proud to host the delivery of the first Boeing 777 and the christening of the Northrop B-2 *Spirit of Washington*.

In 1996, the Museum of Flight began a five-year run of "FlyerWorks," joining the Fourth of Jul-Ivar's celebrations at Myrtle Edwards Park on Seattle's downtown waterfront. The daylong affair included sky parades of antique and military aircraft over the waters of Elliott Bay, with paper-airplane contests and demonstrations on land for the crowds anticipating the annual fireworks display. In partnership with Ivar's Restaurants, the activities entertained hundreds of thousands of spectators in the park and even more watching the television broadcasts at home.

With a continuing buildup of education staff and programs, exhibit improvements occurred over those growth years, with some larger additions requiring millions of dollars of capital investment. The cost of each new exhibit or remodel was estimated, and donors identified and recruited. This series of campaigns fueled the growth in discrete increments.

In 1997, all but one aircraft suspended in the Great Gallery were lowered and cleaned and then rehung in a new pattern. This effort was in preparation for the additions and modifications to the floor-based exhibits and interpretive materials. Former Chief Operating Officer Richard Beckerman organized the task, with support from Century Aviation of Wenatchee, Washington, and a show-business rigger Beckerman had worked with earlier in his career. Tom Cathcart from restoration and the museum security staff assisted in carefully organizing the operation.

Air casters moved the heavier artifacts on the gallery floor. Electric chain hoists were used instead of cranes to lower and raise aircraft, emitting no fumes yet churning plenty of power. The repainting of the DC-3 in Alaska Airlines livery was an exception, and the aircraft was gingerly lowered and surrounded by a Visqueen tent with 20-hp fans to exhaust the fumes.

Top: The Museum of Flight's 247D with the first Boeing 777, both in UAL livery, 1995.

Bottom: Repainting of the DC-3 during the 1997 rehanging of the Great Gallery.

Ultimately, the staff brought down every plane except the Beech C-45 Expeditor, although the team inspected its hanging rigging. This massive job was done quickly and without incident. Concurrently, the 737 fuselage was installed in the Great Gallery lower level as part of the revised exhibit organization. Beckerman remembered this task as one of the best organized and most rewarding of all that he supervised at the museum. Accomplished over seven months, on a tight budget and with the doors open to visitors, this major remodel proved the competence of the museum management.

The Great Gallery had been well engineered for aircraft suspension, yet much of the museum's cost was sunk into the ground. The high water table necessitated protection against intrusion into the basement level, keeping it dry through several flooding events. The complex was built on a fault line, which received particular attention from the engineers in the bracing of the historic Red Barn. Also, cluster pilings were installed as the foundation for the Great Gallery, creating a floating base in case of ground movement. Even with all the design precautions, no one knew how it would react in a quake.

On February 28, 2001, the museum had its test when the magnitude 6.8 Nisqually earthquake shook the Puget Sound region. Boeing Field, built on uncompacted fill materials, was an area of extensive ground acceleration, resulting in considerable destruction. The quake damage to runways, taxiways, and other features caused closure of the airport. The response to repairs took 15 days to allow the airport to open for business.

Earthquake damage in the archives, 2001.

The MOF, notably the Great Gallery of glass and steel, with tons of valuable hardware suspended from its ceiling, demonstrated its exceptional engineering, surviving with minimal damage. Beckerman remembered the shaking that morning. His office was adjacent to the library, then located in the east section of the Great Gallery. The structure performed well, with the library shelving sustaining the most damage. Beckerman took cover under his desk for the duration. "The rows of bookshelves were lined up like dominoes, with stout earthquake straps, and that worked fine," he said. However, the shelves scissored, spilling books out onto the floor. "After getting out from under my desk and stepping over the contents of the file cabinet that had fallen over, I opened my door and threaded through the mess of books."

The Flight Zone opened in July 2001.

Another staff member who watched the gallery from the balcony reported that he was amazed to see all the hanging aircraft swinging in a slight pendulum motion, fortunately in unison, sustaining no damage. Some of the smaller planes continued their sway for several minutes after the quake subsided. Engineering Services Manager Fenton Kraft remembered the shaking interrupting a staff meeting. He looked out to the display berm to see Air Force One rocking with the movement.

With the museum closed to visitors and employees dispersed for the day, Kraft and his facilities management crew searched for potential gas leaks and other problems, finding only minor cosmetic damage—cracks in plaster, hard finishes, and concrete— but no significant structural damage. The museum opened for business the next day.

"The Flight Zone" replaced the award-winning "Hangar" for youths in 2001, occupying the same section of the Great Gallery. With information gleaned over the years, and a significant gift from Microsoft, users, docents, and teachers helped to conceptualize the revised space improvements, which expanded the exhibit into a more immersive experience.

By 2001, the museum's proposal for a National Flight Interpretive Center at Paine Field, initiated in 1990, had transitioned into a Snohomish County-based project that the Museum of Flight still administered. The attempt for partnership with the National Park Service did not materialize, which necessitated a new look at how the complex should be structured and funded. But there was continuing museum commitment to the potential for this important attraction and what it could offer, and in September of that year, the board passed a resolution to continue this expansion.

Paine Field is a working airport with Boeing Company presence. Its plant tours draw visitors from around the world, and the MOF restoration center added another instructional element. Private collectors were beginning to restore and fly vintage aircraft at the field. All of these components added up to a unique opportunity to interpret flight.

Model of the National Flight Interpretive Center at Paine Field.

The Boeing Company expressed interest in moving its factory tour operations into new museum facilities, which was considered the hook for the organization of the components. The museum continued to pursue this opportunity, searching for available airport land and a source for building capital as well as a means for sustaining the National Flight Interpretive Center. Staff and consultants devised several plans for location and design of facilities.

At the time, a Snohomish County Public Facilities District (PFD) gathered tax revenue from the hospitality industry to distribute to projects that spurred tourism. The museum promoted the Paine Field interpretive center as a worthy recipient of these funds for capital and operating support. An additional strategy outlined in this package was to establish a Snohomish County Museum of Flight Authority similar to that in King County. This structure was proposed to provide the legal entity for lease and funding arrangements at Paine Field, comparable to the King County Authority's coordinated support at Boeing Field.

NFIC transitioned into a Snohomish County project early in the new century. The PFD saw value in the project and helped to develop a funding and financing strategy. Officials chartered a new nonprofit, the Future of Flight Foundation, with a nominal land-lease from Snohomish County, as a mechanism for funding, building, and operating the renamed Future of Flight. The Boeing Company partnered with the project and assisted in establishing an orientation center for greeting and preparing visitors for plant tours. This component cinched the new deal. With funding and financing support from the Snohomish County PFD and a capital contribution from Boeing, Future of Flight finished design and went into construction.

FOF opened its new 73,000-square-foot center on December 17, 2005. The $23.5 million facility incorporates an aviation gallery, airport observation deck, and factory tour orientation center, along with offices, a theater, and meeting and event spaces. The partnership with Snohomish County and The Boeing Company has sustained the center, which hosts hundreds of thousands of visitors each year. In conjunction with the MOF restoration center, the additional collections of Flying Heritage and the Historic Flight Foundation, along with frequent warbird flights, have made Paine Field another aviation interpretive center of significance.

Initiated by the Museum of Flight decades earlier, Future of Flight (renamed Institute of Flight in 2015) proves that a good idea will survive setbacks, and can ultimately find a new route. The MOF supported this center throughout its transition, providing aircraft for gallery display. Friends of the MOF put together the investment package for the new hotel for tourist accommodations and catering facilities built adjacent to the center. The MOF continues to loan aircraft. The experience with this project proves that a great institution such as the MOF can spin off resources and influence in many ways.

Champlin's Albatros D.Va (L24) in Mesa, Arizona, prior to transfer to the Museum of Flight.

As the new millennium approached, the museum became aware that the much-envied Champlin Fighter Plane Collection was on the market. For years, museums and individual collectors had salivated at the thought that Doug Champlin's priceless warbirds would become available for sale or loan. Organized over three decades, his restored World War I and World War II aircraft were some of the best. Champlin called it his "legacy," and now he was floating the idea of selling his collection.

His intention to sell was not a secret, nor was the collection hidden, but no one could know what Champlin would do and when. He had moved his aircraft from Oklahoma to Mesa, Arizona, in 1981 to make them more accessible to visitors, but with the changing area demographics and tourism interests, he could no longer continue to operate the museum. When this collection appeared on the market, it was a significant attraction to many.

Ralph Bufano had communicated with Doug Champlin for years and saw this as an extraordinary opportunity that would come along only once in an institution's lifetime. The museum was enthusiastic but also guarded in its expectations. The acquisition of this unparalleled warplane collection was tempting to museums as well as collectors around the globe, as there was nothing else available of this size and historical value. Certainly, the sale would attract a crowd of competitors in a fast-paced bidding process. The acquisition would cost a lot, and a new gallery to house it even more. Master planning had not provided for this expansion, and there were massive challenges in the context of an economic downturn in the Seattle region.

Museum leadership discussed pros and cons and the level of urgency. They decided to move fast to acquire the collection intact so that it would not be sold off in small lots, and then figure out the housing and funding. The decision to unleash a negotiation absent the purchase fund was daring but necessary, risk-taking at its best.

The museum informed Champlin of their interest and began work in earnest to make a deal, to find the money for the acquisition, and then to build a suitable display gallery. All of this was planned for completion on an unparalleled fast track. The newly energized board demonstrated its willingness to seize an opportunity and its ability to meet a big challenge. This achievement is a milestone in the history of the museum.

Trustee Doug Champlin with a P-38 Lightning, part of the collection acquired from him by the museum.

Though departing from the original master plan, this effort set expectations for development in addition to finding valuable land the museum desperately needed. For years, Boeing had declared that none of its corporate properties would be available for the museum. The situation changed when the company moved its headquarters to Chicago and some of their Boeing Field area properties became surplus to need. The availability of additional land parcels from The Boeing Company provided new options for expansion directly to the north on property adjacent to the museum, and also to the west across East Marginal Way. The offer of a donation of land from The Boeing Company, valued at $5.8 million, was an unexpected opportunity that helped advance the Personal Courage Wing.

Initial discussions were held with another warbird collector to negotiate a partnership. In this concept, the partner would buy the collection and put most of the aircraft into a building on the newly acquired property abutting the airfield.

Several designs were proposed in the negotiation, based on a hangar facility for some of the flyable warbirds. These concepts ranged in price, becoming more costly as the negotiations continued. After a year, the partners abandoned discussions, leaving the museum alone in a situation where it needed a new display gallery that it felt it could not afford.

Top: Curtiss P-40N Warhawk prepped for transport from Champlin Fighter Museum to the Museum of Flight, 2000.

Bottom: An expansion model designed prior to the Champlin collection opportunity, with a lid covering the Red Barn.

Everyone knew that this incredible collection was an opportunity not likely to come up again. Champlin's warbirds offered a new order of magnitude for exhibits, an addition that would have a significant influence on attracting a broader audience. The museum had to move ahead in spite of the economic panic of the dot-com setback and the fact that the museum had already been to the well several times.

In November 2000, the museum negotiated with Doug Champlin for a favorable purchase price of $12.5 million. The board financed the loan and organized the capital campaign. Champlin came along with the collection, agreeing to serve the museum as Curator Emeritus. The MOF was off and running on one of its greatest adventures.

Under the rubric "Sky Without Limits," the campaign fulfilled a responsibility to tell the tale of military aviation, focusing on the stories of World War I and II. With the acquisition of the Champlin aircraft, the museum could present an ultimate warbird experience with national implications. This addition raised the museum to a higher level of artifact interpretation and attracted historical organizations to partner and co-locate.

Trustee Bruce McCaw supported fundraising with the offer of his personal aircraft for flying prospective donors to Mesa for a tour of the collection with Champlin. Boarding at Boeing Field, the group made the round trip on the same day. The tour and hospitality served as excellent cultivation for contributions.

The museum and consultants developed a conceptual scheme for expansion, with estimated costs and a schedule. The design incorporated improvements to the entry gallery, enlargement of the lobby and gift shop, and development to both the north and west.

The proposed addition to the north incorporated a gallery for the Champlin collection and a flying hangar. But at this time, there was still priority interest in an expansion to the west, which would require a connector over East Marginal Way. The idea of a grand canopy covering the Red Barn and associated properties was abandoned as too costly, which worked out better aesthetically. The west side was identified to provide a structure to house the large artifacts. Plans were flexible, and a number of options were evaluated.

The museum announced Phase 1 of a revised master plan to build an addition called The Courage Wing in 2001. Concept drawings of the exhibit galleries illustrated military themes and 25 of the warbirds from the Champlin Collection.

In a weak economy, a year with reductions in staff and salaries, the museum hunkered down with donors and produced one of the most significant advancements in museum history.

In 2002, board chair Gene McBrayer professed that this was a time for building—members, structures, exhibits, and passion. This was his call to arms for the institution and its supporters during a time of national concern and economic downturn. Practically, this was not a likely time to build, but if the museum was to seize the opportunity, there was no time to waste.

Tom O'Keefe chaired the campaign for the Personal Courage Wing, with the 60-member board of trustees contributing $20 million.

Top: Concept drawing of a trench scene exhibit for the Personal Courage Wing.

Bottom: In 2003, Harrison Ford narrated and starred in a promotional film for the Museum of Flight during the aviation centennial.

"The Personal Courage Wing is a wonderfully vibrant educational facility," said O'Keefe. "It got me excited because there's such a history there, and I knew it was important to tell the story of the people who not only flew the planes but also worked behind the scenes building and servicing them."

With the start of the "Sky Without Limits" capital campaign and some imaginative land swaps between the museum, The Boeing Company, and the airport, the expansion was secured. Property to the north of the Great Gallery that held T-hangars was identified as the site for the Personal Courage Wing. A proposed Midfield Airpark, a joint project between the museum and King County, was planned as the replacement property for the T-hangars and part of the land-exchange agreement. The museum took down the existing hangars and moved and rebuilt them in designated property on the north part of the airfield.

Top left: Rendering of the massing of the PCW around the north side of the Red Barn.

Top right: King County Executive Ron Sims (right) with Carol Suomi (FAA), Gene McBrayer (MOF), and Bob Watt (Boeing).

Bottom: The Goodyear F2G-1 Super Corsair is loaded for transport to the Museum of Flight, 2004.

Groundbreaking for the Personal Courage Wing was on June 18, 2002, with King County Executive Ron Sims, City of Tukwila Mayor Steve Mullet, and museum trustees and supporters in attendance. The design for the addition, a black box facility, was a dramatic step from all-glass in the Great Gallery to no-glass in the Personal Courage Wing. Rather than a hangar-type facility as first envisioned, this became a highly interpretive gallery with dioramas and stage settings and interactive exhibits.

NBBJ Architects estimated the cost at $47 million, which included 93,000 square feet of improvements. This concept was fluid, as conditions and priorities changed several times. The package included the exhibit gallery and some ancillary spaces, as well as the expansion of the lobby and gift shop.

Sellen Construction served as general contractor, and Seneca Group advised project planning, entitlements, and exhibits contracting. The $53.5 million expansion, completed in August 2003, ultimately provided 88,000 square feet, with two floors of aviation exhibits, a basement that was to find several uses near-term, and an additional 5,000 square feet of flexible space that everyone knew would be rapidly filled.

During construction, Champlin's planes were disassembled and trucked from Arizona to Seattle. Exhibits were crafted for each of the warbirds to include stories of the men and women who designed, built, and flew the aircraft, enriched with artifacts, personal mementos, and multimedia theater.

The Charles Simonyi Fund for Arts and Sciences presented the "Wings of Heroes" gala on the evening of June 5, 2004, including tours of the new wing and cocktails in the Great Gallery. The museum dedicated a tribute to military heroes including Flying Tigers, Tuskegee Airmen, the American Fighter Aces, and Medal of Honor recipients. During the festive evening, one of the honorees proclaimed, "I've never seen anything like this in all the years since World War II."

A "Raise Your Paddle" appeal followed dinner in the gala tent. Frank Sinatra Jr. serenaded the crowd, with dancing to the Harry James Orchestra directed by Fred Radke and special guest Gina Funes.

The museum celebrated the public grand opening of the J. Elroy McCaw Personal Courage Wing (PCW) on June 6, 2004, the 60th anniversary of D-Day, the Allied invasion of German-occupied Europe. General John Shalikashvili, former chair of the Joint Chiefs of Staff, gave the keynote address.

The Champlin Fighter Collection exhibits, presented within dioramas and layered with audiovisual insights and interpretive materials, are rich in content, creating dramatic stories of the aircraft and human achievements. The first-floor galleries highlight the planes and stories of World War II, with the second floor dedicated to World War I. Donald S. Lopez, deputy director of the Smithsonian's NASM, proclaimed, "The Champlin collection makes the Museum of Flight quite a formidable force in the aerospace museum business."

The gallery also wisely incorporated much-needed additions to the museum entrance and lobby as well as gift shop, named the Keith W. McCaw Dream of Flight Lobby. These improvements all contributed to a higher level of hospitality. Visitors were better received and ticketed, and the store could sell more merchandise. The range of meetings and banquets was expanded. This combination created a superior museum, as was immediately reflected in the crowds of visitors and increase in operating income.

The Personal Courage Wing effectively added a new component of interpretation and created renewed excitement for the museum. Fresh audiences were attracted, and previous visitors, now more than ever, had reasons to return. To quote Alan Shepard after his return from the successful Apollo 14 flight, "It's been a long way, but we're here."

With the success of the Personal Courage Wing, the museum returned to planning in 2004. Although there was progress, substantial projects continued to need attention and organization for capital funding. Ralph Bufano and chair Bruce McCaw asked the committees to move all of the required components back on track.

Top: The Personal Courage Wing rises above the Red Barn to the north.

Top: The FG-1D Corsair is displayed with its wings folded on a simulated aircraft carrier flight deck in the Personal Courage Wing.

Bottom: Fokker Dr. 1 replica soars in the World War I gallery of the Personal Courage Wing.

Planning at this time identified three ad hoc committees to advance development. Bufano and McCaw accepted leadership roles on all. The Pavilion Committee, working on what at that time was referred to as the Personal Courage Wing addition, was to address the design and building costs of a structure to incorporate room for an Education Center and Library and Archives with Collection Storage. As planning moved forward, this was not to be. The Commercial Transport Wing Committee considered a structure for the museum's largest artifacts, to incorporate space for additional collection storage. The Development Feasibility Committee was tasked with assessing the potential of these projects and options for funding them.

With new opportunities, the foundation proved its ability to flex, and the work began.

Bruce R. McCaw

For the past 35 years, Bruce R. McCaw has set the standard as one of the Museum of Flight's most loyal and generous supporters. His penchant for aviation combined with his willingness to volunteer and share his wisdom, resources, and connections created a partnership that benefits all aspects of the museum from planning and building to operations.

Bruce, the eldest of four brothers, was born in 1946 in Washington, D.C., where his father was stationed at the Pentagon. A pioneer broadcaster before the war, his father, J. Elroy, served in World War II as a communications specialist and continued his passion afterward by establishing radio, television, and cable TV companies from Hawaii to New York. His mother, Marion, one of the first female accounting graduates at the University of Washington, was also a business person and supporter of the arts, particularly opera. Bruce, along with his brothers, attended Lakeside School. All held positions in their family's businesses and continued on to higher education, followed by many ventures, both together and separately.

After their father's death in 1969 they banded together to expand a tiny cable television system in Centralia, Washington, and built it into the 20th-largest in the nation. Subsequently, they became innovators in the burgeoning mobile telephone industry, creating McCaw Cellular Communications Inc. In 1994 AT&T acquired their company, becoming AT&T Wireless.

Equal to his love of aviation is Bruce's passion for motorcars and racing. As a child, he was captivated with driving while sitting on his father's lap in the family automobile. McCaw tells entertaining stories of his youthful exploits speeding cars and motorcycles through his and friends' neighborhoods. "By the time I was 12 or 13, I was a pretty decent driver, and I could hustle a car down the road." He has never lost the passion for speed.

An avid collector of classic motorcars, vintage racers, and automotive memorabilia, Bruce has continued to participate behind the wheel at the annual Pacific Northwest Historics fundraising races. He is well known in the motorsports community as a driver, team owner, and collector, and has won competitions in all categories. Bruce was a director of Championship Auto Racing Teams (CART) and founder and CEO of PacWest Racing and subsidiaries.

Bruce's love of racing machines is not limited to cars and airplanes. His restoration of the historic Slo-Mo-Shun V hydroplane has been a pet project. The legendary speedboat won the very first Unlimited race in Seattle, the 1951 Gold Cup. It would later win two additional Gold Cup races before its famous flip in 1955, on the same weekend as the equally famous "Gold Cup Roll" of the Boeing 707 "Dash 80" prototype over the hydroplane course.

A pilot for more than 40 years, Bruce holds commercial and instrument ratings in Learjet, Falcon 50/900, and F-27 aircraft, and is a member of Quiet Birdmen, the Wings Club, NAA, and NBAA. He is a recipient of the Living Legends of Aviation Award.

In 1981, Bruce cofounded Horizon Air and served as vice president and director of the company. Headquartered in Seattle, it initially flew three Fairchild F-27 aircraft from Seattle to Yakima and Pasco, Washington, as well as Sun Valley, Idaho. The fledgling airline acquired Air Oregon and Transwestern Airlines and expanded rapidly into five states and 30 cities prior to Alaska Airlines' purchase of the company in 1986.

In 1988 Bruce, along with Museum of Flight trustees Clay Lacy and Joe Clark, organized Friendship One, an around-the-world flight in a Boeing 747 to raise funds for children's charities. The plane and its passengers set the record at 36 hours, 54 minutes, and 15 seconds, raising more than $500,000.

In an out-of-world experience, Bruce was a guest of the Soviet delegation on the flight of an aging Ilyushin IL-62 from Seattle to Moscow in 1991. Cosmonaut Igor Volk was the flight chief, and in the midst of an epic onboard party, he decided to divert the flight to Petropavlovsk, a top-secret Soviet Air Force and Navy base, closed to all but the highest-level Soviets and off-limits to any Americans. While the flight crew rotated between flying and drinking and singing, McCaw inherited the vital role of communicating with a flabbergasted U.S. air controller. Volk stubbornly pushed toward the military port city. Although unauthorized, it was the first American visit to the base since WWII. McCaw and Volk and the rest of the passengers and crew did not miss a trick, partying until daybreak. The few Americans on board that historic flight will never forget it.

Bruce McCaw is currently co-chair of the Apex Foundation, chairman of Seattle Hotel Group, and chairman emeritus of Pistol Creek Company. He was a cofounder and director of Claircom Communications, chairman and founder of Forbes Westar, Inc., and a director of Alaska Air Group.

McCaw is active in numerous charitable institutions, including the Friendship Foundation. He is currently a trustee of Behind the Badge Foundation, Seattle Opera, Congressional Medal of Honor Foundation, Museum of Flight, and the National Air and Space Museum. In the past, he has served on the boards of Lakeside School, St. Thomas School, Talaris Institute, PONCHO, and WSU Foundation.

Bruce's contributions of service to the community are recognized in the Puget Sound region and beyond. In 2004, he and his family received Seattle's First Citizen Award. The Washington Policy Center awarded Bruce the Stanley O. McNaughton Champion of Freedom Award in 2007. In that same year, the Northwest Chapter of the National Association of Fundraising Professionals honored his Apex Foundation as Outstanding Philanthropic Foundation of the year. In 2009, Bruce was awarded the Woodrow Wilson Award and the Behind the Badge Law Enforcement

Bruce in the left seat of the Learjet 45 first delivery, 1998.

Citizen of the Year Award. Lakeside School presented Bruce with the Willard J. Wright Distinguished Service Award in 2008. In 2015, Colorado College presented Bruce with a "Doctor of Humane Letters, *honoris causa*." He is the proud father of three children.

During his tenure as a trustee of the Museum of Flight, Bruce has been a long-standing member of the Executive Committee and served on nearly every other committee with a two-year term as chair. The museum rewarded his contributions to aviation with a prestigious Pathfinder Award in 2013. Although his philanthropic interests extend to art and history organizations, youth charities and education across the nation, his spirit is ingrained at the Museum of Flight and his impact indelible.

Left: The British Airways Concorde arrives on its final flight, November 5, 2003.

Right: The museum's Wright Flyer replica was featured in the *Birth of Aviation* exhibit.

CHAPTER FIFTEEN

20th-Century Treasures in the 21st Century

The beginning of the 21st century brought flight treasures large and small to the museum, dramatic additions that set a new level of attraction. The Centennial of Flight fueled the next stage of growth.

The year 2003 was the epochal century celebration of the first powered, controlled, heavier-than-air flight and a significant milestone in American history. The MOF seized on this event and produced what was heralded as the most comprehensive exhibit about the Wrights and their accomplishments anywhere in the country. Incorporated into the *Birth of Aviation* exhibit were the Wright Papers, never-before-seen documentation of the work of the Wright brothers.

The Wright documents, long thought lost or destroyed, were discovered in a private collection, and the museum purchased the papers with a gift from a board member. This acquisition, one of the most valuable in the extensive archives, was a fascinating adventure of opportunity and response, another occasion that belies the stereotype of a dusty museum.

These papers were out of sight for years and expanded with the affiliations of both the Martin and Curtiss companies. Orville Wright sold all aviation patents and materials to a New York syndicate in 1915, resulting a year later in the formation of the Wright-Martin Company. Then, in 1929 the company merged with Curtiss to form Curtiss-Wright, and all the accumulated papers moved along with the corporate archives.

When Curtiss-Wright closed suddenly in 1984, all of these records were judged surplus. An employee tasked with disposing of these materials sensed they were relevant and retrieved them from the dumpster. The documents were temporarily saved, though still at risk. The next step involved a father-and-son team of collectors.

Joe Gertler Sr. assembled one of the largest stockpiles of race car and aero artifacts in the country. Gertler Sr. was a craftsman with an interest in speed sports, race cars, boats, and aircraft. Operating out of his Bronx Raceway Garage, he purchased and salvaged full-sized models and parts, organizing what he called The Raceway Collection in 1939. His son, Joe Jr., joined him in 1969, and their growing inventory of vintage racing cars, aircraft, engines, and related items caused them to move to new quarters for storage.

Gertler Sr. died in 1990 and Junior continued the business, relocating in 1993 to Florida. Needing to pare down the collection for the move, he held an auction. Though it was well received, he still had five large tractor-trailers full of materials. Gertler Jr. began to focus on library and archival materials and art, and he is still in business purchasing, trading, and selling.

Before his move south, Gertler Jr. received a call from a man who claimed to have the corporate records of the Curtiss-Wright Corporation that contained the charter of the Wright Company. Gertler thought this was just another of the many crank calls he experiences, but he did follow up with the man, who was a former Curtiss-Wright employee.

This 1909 document of incorporation of the Wright Company, signed by Orville and Wilbur Wright, is an important part of the museum's collection.

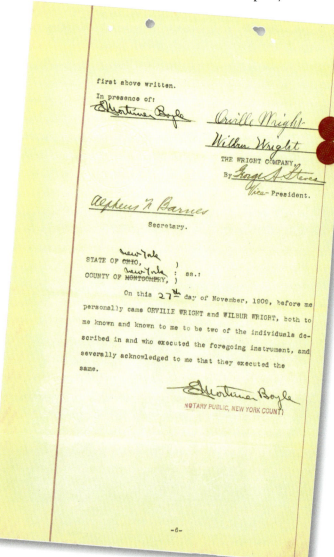

Gertler was amazed to find that this was most probably the real deal and began several years of conversations before finally striking an agreement to purchase the boxes of materials in 1992. In addition to the Wright papers were archival documents of the Curtiss Aeroplane and Motor Company and the Glenn L. Martin Company, which merged over the years. These corporate archives offer an overview of early American aviation. Gertler Jr. did not want to sell the materials piecemeal, though that might have fetched him the most money, so he began searching for a source that would buy the entire lot. There was interest, but the funds to purchase were not readily available.

MOF trustee Jon Shirley was visiting Gertler in Florida in search of an automobile part he needed and learned of the Curtiss-Wright papers. Shirley brought the find to the attention of the museum, calling Bruce McCaw with the news that this was something worth pursuing. McCaw concurred.

In 1997, MOF curator Dennis Parks flew to Florida to check out the materials and said: "I was flabbergasted. I was looking at the birth certificate of the beginning of aviation manufacturing."

Parks authenticated the collection, appraised at $1 million, and he and Ralph Bufano began negotiations. McCaw stepped in and through his charitable foundations provided the $1 million it took to purchase this incredible set of documents. The museum structured a two-year payout, with some materials coming as a donation for tax purposes to offset the portion purchased. This deal was good for all parties, and the museum again was instrumental in pulling a priceless treasure from a dumpster.

After these precious papers were acquired in 2002, their presence was kept relatively quiet while they underwent organization and conservation. The documents provided an unparalleled resource for the centennial celebration. In the words of Parks, "These papers—which include legal documents, correspondence, meeting minutes, and much more . . . positioned the MOF to tell the story of how the Wrights turned an invention into an industry more completely than anyone ever has before."

Curator Dennis Parks, shown here with the museum's 1914 Caproni CA.20, knew the real significance of the Wright papers for aviation history.

The Wright brothers were close partners in a stable family who also enjoyed vital support from their sister Katharine. The narrative of their genius, hard work, and spirit is an uplifting American story. The challenge for the *Birth of Aviation* exhibit was to overcome stereotypes of the Wrights and their enterprise. The brothers were more than just eccentric bicycle mechanics and tinkerers; they were determined engineers, researchers, and explorers. Some scholarly books have advanced

Artist's rendering of the *Birth of Aviation* exhibit, 2003.

this theme, but the general public was not aware. The importance of this message intensified the museum production and set a hallmark for its comprehensive presentations.

The exhibition was curated and designed in-house and enriched with programs and activities for young and old and scholars. In addition to revealing their new collection, the museum gathered loaned documents and featured an excellent display of full-scale early glider replicas.

Top: 1902 Wright Glider replica.

Bottom: Visitors at the *Birth of Aviation* exhibit, 2003.

Marketed around the country, *Birth of Aviation* ran for six months, from August 2003 until February 2004, and was a major smash. In concert with the show, the museum was the only West Coast venue for the Experimental Aircraft Association's exhibit *Countdown to Kitty Hawk*, which traveled nationally with a carefully replicated Wright Flyer that was scheduled for a re-creation flight at Kitty Hawk on the upcoming 100th anniversary. This aircraft was on display from show opening until September, allowing close-up views of the frail machine that ushered in powered flight. Later, in 2005, with assistance from Bruce McCaw, the museum purchased this replica for permanent display.

This 5,000-square-foot blockbuster was installed in the Personal Courage Wing (PCW) before its grand opening. There was something unique and fitting about this exhibit, serving as an introduction to the new gallery. In the words of CEO Bufano, "It is entirely appropriate that the *Birth of Aviation* is the inaugural exhibit in the Personal Courage Wing, because the story of the Wrights' triumph is very much one of personal courage: courage in the face of discouragement, ridicule, physical danger, and the ever-present risk of failure." This grand story was worth telling, and the museum found the content to tell it well. Some still look at this period as one where the institution put on its long pants.

Museum Soul

The paper trail for the library and archives is another story of opportunity and adventure. Lovering recalled someone depicting an archive as "the soul of the museum." A library lover who had worked his way through the University of Washington employed in the stacks of the Suzzallo Library, Lovering picked up on the idea, though he was not the first.

Museum founder Harl Brackin was determined to build an archive. As a corporate historian at The Boeing Company, Brackin was troubled by the loss of documentation. Many others among leadership focused upon major artifacts, principally aircraft, but Brackin was steadfast in saving the paper, the plans, the technical manuals, and the photographs.

The Museum of Flight archival collection contains both paper-based and photographic materials. The documents span aerospace from the Wright brothers to space exploration.

The museum possesses millions of rare photographic prints, negatives, and slides that include images of aircraft, aerospace personalities, events, and places of historical importance to aviation and space exploration. The paper-based items include primary source materials such as personal files, clippings, documentation, maps and charts, airline schedules, ephemera, and drawings.

Housed now in a building across the street from the main complex, the Museum of Flight Research Center includes the Museum of Flight Archives, the Harl V. Brackin Jr. Memorial Library, and the Kenneth H. Dahlberg Military Aviation Research Center. The Brackin Library contains thousands of books, magazines, and manuals, with an information system with more than 300,000 references. The Kenneth H. Dahlberg Center covers military aviation history, including materials from the American Fighter Aces Association, American Volunteer Group, and the Champlin Fighter Museum, as well as the expansive Norm Taylor and A. J. Bibbee photograph collections.

Dennis Parks was the fourth curator at the MOF, serving from 1996 to 2008. He worked on many of the valuable additions to the archival collection. As Curator Emeritus, Parks still comes in as a volunteer and to get material for articles he writes.

Parks believes that among the most valuable is the vast Peter M. Bowers Collection, containing more than 250,000 prints and negatives that are a complete overview of all aircraft ever built. These include 100,000 photos that date from before World War I. Bowers had a profound effect upon future aviation professionals of every description through his photographs of the evolution of flight history.

Pete Bowers was a member of what was called the "616 Photo Club" (for the large-format film, rectangular and suited to aircraft). He had taken photographs of aircraft and written about aviation from his high school days. Pete had a prolific publishing career, amassing a valuable collection of photos. He was a prolific writer as well as a photographer, and his photos have appeared in many journals and books.

The Kenneth H. Dahlberg Center.

The David D. Hatfield Collection documents the full history of aviation in California. Hatfield was in real estate, but his passion was aviation, and he gathered everything he could, including magazines, papers, drawings, books, and scrapbooks of articles and photos. Every flight institution wanted what Hatfield amassed over many decades.

David Price, the founder of the Museum of Flying in Santa Monica, California, had control of the Hatfield Collection and hoped to make it part of his museum. This did not come to pass, and Price decided to sell the collection. This treasure trove consists of 5,000 books, scrapbooks of photos and articles, as well as 1,000 8-by-10-inch photos of women in aviation and 30,000 to 40,000 negatives of flight subjects dating from before World War I.

In the deal, Price included 1,000 albums from the Douglas Library, containing original Douglas Aircraft Company drawings of the World Cruiser, Cloudster, and other aircraft. The Museum of Flight's winning bid incorporated a trade of a 727 Cockpit Procedures Trainer. All of these drawings were very fragile, so Parks sought and received a grant to digitize the information.

Among the most precious of acquisitions were the Wright Aeronautical Corporation papers covering the incorporation of the first aircraft company in the United States. Parks mentioned that this was a find with a fascinating back story. This addition is an odyssey from the Wrights to Martin, to Curtiss, to a fortunate save, to collector Gertler, then finally to the museum. This was another fantastic capture of fortune from a dumpster.

Boeing Corporate records joined Douglas Corporation materials upon the merger of Boeing and McDonnell Douglas. These records include payroll documents on employees who represent a who's who list of aviation leaders in those early days.

The Elrey B. Jeppesen Collection contains the personal and professional memorabilia of the creator of the Jeppesen aerial navigation charts, including his famous "Little Black Book." These materials also came along partially as a result of Boeing purchasing the Jeppesen Company. Other museums had been after this collection, principally the little black book of flight maps compiled by Jepp. MOF chased this for years, as did others. Jepp personally visited the MOF and was extremely pleased with the facility and archive, and selected the museum as the recipient of his papers.

Ned Thorne, a longtime volunteer, added his German aviation photos from pre-WWI through WWII. Thorne had worked at Wright-Patterson AFB during WWII and assembled these valuable prints as they were surplus at the base. Other materials were targeted for acquisition. The Norm Taylor photos added one of the largest private military aviation collections.

Gordon S. Williams, a cohort of Peter M. Bowers, began photographing aircraft as a teenager and worked professionally as a Boeing Company photographer. His photographic materials contain more than 100,000 negatives depicting aircraft from the 1930s through the 1980s.

The Lear Collection contains drawings, photographs, and documents on the history of the Lear Aircraft Corporation. It is another family treasure that came to the MOF because of the involvement of Moya Lear, a museum trustee, and her friendship with Bruce McCaw, Joe Clark, Clay Lacy, and Chuck Lyford. Knowing that the museum would take care of the materials, Lear moved them from the University of Texas.

Also, there are valuable documents in the list provided online for the museum archives. These include the Dale Parker Space Collection (Mercury, Gemini, Apollo, Space Shuttle, and Soviet space programs); James McDonnell Collection, with biographical and early McDonnell Aircraft Corporation materials; *Enola Gay* crew oral interviews; George Schairer Engineering Collection,

Grand opening of the Kenneth H. Dahlberg Military Aviation Research Center, 2007. Left to right: Phil Johnson, trustee; Cory Graff, Assistant Curator; George Chandler, president of the American Fighter Aces Association; Kenneth Dahlberg, American Fighter Ace and donor; and Dr. Bonnie J. Dunbar, President & CEO.

dealing with the swept-wing design in Germany; the Wright brothers' 15 glass-plate negatives of Wilbur and the 1908 flights in France; the Jack Kerr slides of the Curtiss-Wright Buffalo, New York, plant in the early 1940s; and the Robert W. Stevens photos documenting aviation history in Alaska.

The staff and volunteers have done a lot to organize, preserve, and record this vast array of records. Parks is particularly proud of the amount of digitizing that staff has accomplished that allows these documents to be shared and used in research and artifact restoration.

The MOF has one of the best and largest air and space history collections in the country, and it continues to grow and expand. It seems as if archival materials were just waiting for a great museum to come along. The aerospace library and archives are an accurate measure of the true greatness of the Museum of Flight.

Also in 2003, the museum prevailed in a global competition for one of a limited number of the iconic Concorde aircraft. At the time of the Concorde's last visit, in 1989, Lovering had made a formal request and was tentatively promised an aircraft if the museum would save space. There was no timetable then, but the museum had been consistent in its petition for a Concorde from at least the time of the first Seattle visit, in 1984.

Executives, up to and including the CEO of The Boeing Company, wrote to request a contribution of the aircraft. The Concorde's two visits to Seattle had cemented relations, but the museum knew it was in a contest. The line of applicants for one of the aircraft was long; many were in North America and even more in Europe, which had a historical advantage.

Former COO Richard Beckerman was on the museum team that developed a formal proposal addressing all the relevant issues and reminding British Airways (BA) of their past partnerships and requests over several decades. The museum wisely portrayed itself as a longtime ardent suitor, not a current opportunist. BA representatives made a site visit to confirm facilities and practices and seemed impressed by the museum's technical proposal to use onboard ducting for a dedicated HVAC system, thereby protecting the aircraft without disturbing its integrity. This engineering technique proved a commitment to conservation initiated with Air Force One, and the museum used it to win other artifact competitions.

Bufano traveled to review the available aircraft from the fleet, identifying as a priority the one that was the last to come out of service. That Concorde was the model gifted to the Museum of Flight after decades of cultivation with BA and its beloved brand. It was a major acquisition, and it was important to the image of the museum. The award announcement was a proud moment.

On the afternoon of November 5, 2003, the Concorde once again arrived at Boeing Field, for the last time. It came home to take up residence as only one of three in the United States, and the only supersonic transport on the West Coast.

A thousand attendees and a gaggle of media responded to the reception invitation, making it clear that no one wanted to miss this historic occasion. With cars still patiently queuing up for valet parking, and lines of enthusiastic partiers picking up their badges, the great swoosh of the British Airways Concorde resounded throughout the museum. Chief Concorde pilot Mike Bannister flew a low pass to the south of the runway, blasted up and out, turned, circled back, and prepared for landing. The Concorde was a bit early for its homecoming.

The excited guests ran to viewing points on both levels of the museum, and the show began. Although the premature arrival toppled the sequence of the reception procession, no one in attendance had any complaint.

The Concorde lands at Boeing Field to become a permanent part of the Museum of Flight collection.

The palpable drama of a Concorde landing trumped all other trappings, and the crowds with binoculars and cameras gathered to watch its slow procession along the taxiway. A bagpipe band and the shouts and cheers of the masses greeted the Concorde.

As in past visits, traffic slowed to a near halt on Interstate 5, and visitors gathered along the fence line of the airport to once again welcome this majestic bird. The aircraft in BA colors is a beautiful sight, and as it approached its place of honor at the museum, the atmosphere was electric. Suddenly, flags poked out from both cockpit windows, with co-pilot Captain Les Brodie

Left: Concorde arrival guests linger to admire the SST, positioned prominently on the museum aircraft ramp.

Bottom: Bruce McCaw (left), Captain Bannister, and Ralph Bufano at the Concorde celebration.

waving the Stars and Stripes, and pilot Captain Mike Bannister brandishing the Union Jack. Those who were there will forever remember this spectacular occasion.

As the crew and guests deplaned to the red carpet, they were met by Washington State Governor Gary Locke, King County Executive Ron Sims, County Council member Pete von Reichbauer, Seattle City Council member-elect Jean Godden, and museum representatives. Upon acceptance of the keys to the aircraft, Locke said, "Concorde is more than just an airplane; it embodies the spirit of adventure in all of us."

The museum now officially had its long-desired Concorde. Over a span of 20 years, the museum had twice brought in the British Airways Concorde for events, hoping someday to acquire one for its collection. November 5 was that day, and it was time for a party.

British Airways and museum representatives joined the throng in the Great Gallery to the tunes of The Nowhere Men, a Beatles cover band, and the celebration kicked into gear. The hard memory that The Boeing Company might have, could have, had a superior model was suspended. After all, this was the final North American flight of any Concorde, and its landing point was the Museum of Flight.

Over the din of the revelers, museum chair Bruce McCaw took the stage in the gallery with chief pilot Mike Bannister and airline executives to thank everyone and place some background on the evening. It was a bittersweet moment for the Concorde crew, saying goodbye to their longtime friend. Captain Bannister said, "There are three great lady loves in my life. My wife, Christine, my nine-year-old daughter, Amy, and Concorde. There's only one of them I can

Aviation Learning Center, 2005.

control. You'll be pleased to know it's that wonderful lady behind me."

The Concorde is one of the few aircraft that is almost as beautiful grounded as in flight, and its very image outside the glass panels thrilled the crowd and ruled the news broadcasts that evening and for days to come.

The promotion "You Will Become a Witness to History" was not hyperbole to those who observed its arrival. The Concorde had made its mark in Seattle and helped the fledgling museum to make an imprint in 1984. This fantastic supersonic aircraft was a treasure desired, competed for, and successfully acquired. It meant a great deal at the moment and would for the future.

The Challenger Learning Center that opened in the Great Gallery had kicked off the museum's entry into space education, and it was evident that similar instruction on aviation was a necessary counterpart. Education Director Bonnie Hilory, who had been on staff in various assignments from 1998, resigned her position to take on the task of building a new type of aviation learning center.

Hilory spent most of a year researching what was available, traveling around the country to look at examples. At the time, NASA had a program called "Castles in the Sky," which Hilory found inadequate. NASA also had a traveling interpretation of space travel that came to the museum with hopes of getting national traction, and Hilory studied the possibility of an MOF partnership. Ultimately, the Museum of Flight decided to take on the mission to imagine, design, and build a center on the same level of quality and popularity as the Challenger.

Concorde

The Anglo-French Concorde is both an inspiring tribute to human ingenuity and a cautionary tale for the global aerospace industry. Placed into service in 1976, this sleek beauty—the world's first supersonic transport (SST)—cruised at just over twice the speed of sound far above other jets. It cut flight times in half, whisking travelers from New York to Paris in just three and a half hours.

By the same token, the Concorde had an earsplittingly loud takeoff, generated sonic booms that precluded supersonic cruise over land, and consumed as much fuel carrying a hundred or so passengers across the North Atlantic as a 747 did with three or four times that many. If it was a technological marvel, it was also an economic white elephant.

From aviation's earliest days, human prowess aloft followed the mantra "higher, faster, farther." World War II greatly accelerated the development of flight technologies, giving the world jet engines and swept wings. The sound barrier fell in 1947, and Mach 2, or twice the speed of sound, was achieved in 1953. Amid this progress, an expectation reigned supreme that airliners too would fly faster than sound.

The British and French had each begun independent SST developments in the 1950s. By the start of the next decade, it appeared that program costs might be prohibitive, so they joined forces under the name Concorde. The first prototype flew at Toulouse, France, in March 1969, and the second in Bristol, England, the following month. Both nations' prototypes attended the 1969 Paris Air Show, where they found themselves unexpectedly dwarfed by the huge and equally new Boeing 747.

At that time, 17 airlines around the world—including seven U.S. carriers—had placed orders for more than a hundred Concordes. But escalating program costs, the dramatic crash of the rival Tupolev Tu-144 at the 1973 Paris Air Show, soaring jet fuel costs in the wake of the Arab Oil Embargo, prohibitions against supersonic flight over land, and the amazing economics of the 747 led to massive cancellations. Ultimately, just 20 Concordes were built and only 14 of them ever entered service, seven with each sponsoring nation's flag carrier.

G-BOAG stopped traffic on I-5 when it landed at Boeing Field.

British Airways and Air France garnered prestige with their premium-class Concorde services, which catered to the very affluent and those seeking a unique experience. Displays on the forward bulkhead told those aboard how fast and high they were flying as well as the outside air temperature. Concorde services continued until both airlines retired their fleets in 2003.

The Museum of Flight's beautiful example, G-BOAG, flew into Seattle in November 2003. Visitors who tour this airplane's cabin are generally amazed at how cramped it feels inside. Although clad in elegant leather, the seats are small. So too are the windows and very narrow aisle, while the ceiling and doors are both quite low. Observations of this nature entirely miss the point—so fast was the Concorde that passengers had time to feel thrilled but never uncomfortable.

The original theme for the Aviation Learning Center (ALC) was "day in the life of a pilot." Hilory wrote a prospectus, organized a team, hired consultants, and went to work. Mike Koss of Microsoft, who was involved with the application Excel, became a prominent resource for coding of simulated programs, later joining the museum board. Other Microsoft techies, most from the Flight Simulator division, joined up as volunteers. Hilory's "Tiger Teams" consisted of an education staffer, volunteers with particular expertise, and a consultant on design or fabrication that involved the ALC curriculum, from flying to ground-school training to wind tunnels.

With space assigned in the Great Gallery, the ALC cost $2 million; approximately $1.2 million was cash, and the rest contributed items and services. The support from Microsoft—hardware, software, and consultation—was essential to building this learning center. Cirrus Design Corp. donated an airworthy prototype of its SR20 single-engine plane, which profoundly enhanced the flight training sessions. Mahlum Architects of Seattle designed the center, with Dillon Works of Mukilteo, Washington, detailing the exhibit fabrication.

Top: At the ribbon-cutting of the Aviation Learning Center, Harold Carr, Education Chair, is assisted by trustee Bruce McCaw and President and CEO Ralph Bufano.

Bottom: Hands-on learning in the ALC with the Cirrus SR20 in the background.

The ALC opened in the Great Gallery in 2005 as one of the finest hands-on aviation simulation experiences at any museum in the country. It has won numerous awards and has fascinated and motivated hundreds of thousands of children in the years it has been in operation. It is also an excellent teaching platform for the museum's professional education specialists. Combined with the complementary curriculum of ALC, Challenger has become even more valuable and essential, the counterparts enriching one another. At this point, the MOF had moved beyond just a committed education institution into a role as the leader in the transportation museum world.

As is typical of the aerospace economy, during the first decade of the new millennium, the Museum of Flight would experience economic downturns that challenged all nonprofits, particularly those caught in the midst of a capital acquisition campaign. These several recessions, national in scope, one of which was particularly painful in the Puget Sound region, slowed but did not stop the expansion curve.

In the early 2000s and later in the major recession of 2008, the museum looked for ways to maintain essential services within budget reductions. Staff, one of the largest expenses, had to be reduced, which was painful. These modifications were made. The remaining staff accepted reduced salaries with no increases for the next few years, in large part due to their dedication to the mission and loyalty to the institution.

What is most instructive in all of these upticks and downturns is that the museum board found a way to normalize the wild swings and continue the pace of development. Some essential additions occurred in the worst of economic times. Bold decisions, even risky moves, were made to move the museum to a new level of public service. For those seeking an adventure, this might be one that epitomizes the museum and its rapid development.

By the end of his tenure as CEO in 2004, Ralph Bufano could reflect upon his fast-paced experience and take pride in the fact that the Museum of Flight was the foremost educational air and space museum in the world. Not just in the region or country, but in the world. The aggressive objective the board set years earlier had met and exceeded the expectations of many. The museum was a functioning team, highly successful, the envy of any nonprofit.

In response to opportunities, physical facilities had expanded well beyond initial plans. The additional properties enabled the museum to meet visitor demand and house the considerable additions of artifacts and archives. Visitation increased, earned income rose, capital funding relied upon a much broader base, the collection grew to be one of the best of any flight museum, and the exhibits proved captivating. Most important, the institution was surging in its educational mission, serving huge numbers of young and old and defining what an aerospace museum can mean in service to the community.

The MOF was not only an avowed leader among flight museums and the transportation museum sector but also highly recognized by the museum world as a whole. This rush of expansion had taken the foundation to the next stage in its flight, positioning it on a path that showed promise to reach places where none had gone before.

Ralph A. Bufano
President and CEO, the Museum of Flight, 1992-2005

President and CEO
Ralph Bufano
with the museum's
G-BOAG Concorde.

Looking back at my tenure at the Museum of Flight, I think of all the challenges that we all faced together and the many successes we achieved. The successes were because of the people who helped make it all possible, and without the support I had from the trustees, staff, volunteers, donors, and members, I would not have been as successful. I must add here that The Boeing Company and its employees also helped more than I can say.

As president you captain the ship, you set the course and you guide the museum, you look out for trouble ahead and you take advantage of a fair wind, and I must say I enjoyed it all.

I had the advantage of closely surveying the Museum of Flight years before I took the helm as CEO. I was part of the American Association of Museums on-site review for the museum's accreditation team and met with the board of trustees and staff. I was impressed with everyone.

One of the interesting things I learned was from a Boeing employee I met on the plane going to Seattle. He told me about the "Boeing Museum," and I started to correct him that it wasn't the Boeing company museum, and he told me that it was started by the Boeing employees, that it was their museum, and that they were all members. I then understood the community that supported the museum.

I was fortunate to have a strong board and a good staff to start with, and in the first months, I made a primary list of the museum's strengths and weaknesses. The next 15 years were spent working on that list. One of the first initiatives was to build the long-awaited Phase 3 of the museum to house the restaurant and meeting rooms. We didn't have the money, so we borrowed it and paid it back with the new income from catering.

The museum suffered through the dot-com crash and following recession with the rest of the nation, but it gave us a chance to step back and look at what we did well and what was important to us.

In 1999, we held a board of trustees retreat to help us look at our mission and goals for the next decade. At this meeting, we reinforced our commitment to the vision that set us apart from every other museum:

To be the foremost educational air and space museum in the world.

The museum also continued the long-term effort to build a new site at Paine Field for both museum and Boeing tours. Working with the Snohomish County Council and Paine Field Airport staff, we designed and oversaw the building of the now Future of Flight Aviation Center.

The museum followed through with plans to buy the Champlin Fighter Collection and build the Personal Courage Wing. This was no small endeavor. The size of the board grew and so did the collection. As the attendance numbers expanded, we added the new lobby and evaluated the collections and exhibition plans. After looking at the future goals and the financial impact that meeting these goals would require, the board stood tall and supported our forward direction.

Another major accomplishment was our partnership with the Highline School District. When we were approached about the possibility of partnering with them to build an aviation high school, the timing was difficult for us. We were in the middle of several projects, but the board and staff knew that this was the right thing to do. We decided to move forward.

I look back today and know how fortunate the museum has been to have such committed supporters and volunteers.

I believe that the Museum of Flight is a truly great institution and that it will play a national role in guiding America's youth, inspiring them to share the dream of flight, helping them understand our history, motivating them to expand their minds, and helping them create goals of their own.

— Ralph A. Bufano, FRAeS

Left: The North
American P-51D
arrives at the museum
after restoration, 2010.

Right: CEO Bonnie
Dunbar in the
museum's Challenger
Learning Center, 2005.

CHAPTER SIXTEEN

Faster, Higher, Farther

Upon the retirement of Ralph Bufano in 2004, the board selected astronaut and Pathfinder Award honoree Dr. Bonnie Dunbar as the new CEO. Dunbar's service as a Museum of Flight trustee had acquainted her with the organization and operations, allowing her to step in at once and work with the new chair, Jim Johnson, and the board.

Dunbar began work in a period of renewed growth in museum visitors and program participants with a bullish economy. There was movement in all sectors of museum operations, plans for expansion, and additions to the collection. In this rapid pace, she managed to focus attention on the important business of enhancing the educational programming. She had an initial period of fair winds before being buffeted by one of the worst economic recessions in history.

With an astronaut at the controls, the institution enhanced its space interpretation and added muscle with a range of initiatives. Already a recognized leader in youth education and a motivator in encouraging science and technology careers, the museum focused on the opportunity to move from being an informal resource to a partner in promoting Washington State's learning standards.

In her first year, Dunbar reported 474,000 paid visits, now approaching a target set earlier in operations. Earned income had increased with the popularity of the Personal Courage Wing and the expansion of public programs. Dunbar was able to report a 4-star rating from the nonprofit evaluator Charity Navigator in 2005 and just a year later the same high ranking for financial stability from GuideStar. These are valued independent assessments that communicate to donors and supporters the financial position and practices of nonprofit institutions.

Let's Do the Numbers

Upon the opening of the Personal Courage Wing, the museum posted improved numbers all around. Steering out of a bad economy and before being hit head-on by a recession, the discipline in operations and successful expansion of facilities enriched performance and community service.

The museum welcomed 438,000 visitors in 2006 and educated 120,000 students while building annual memberships to 21,000. Earned income from admissions, membership fees, events, and sales increased by 5%, providing 66% of revenues to the $16.5 million budget. The institutional assets grew to $96 million.

This performance demonstrates that in a tough period, the operations were able to adjust mission to budget and upgraded the quality and significance of exhibits and education programs. As well, the foundation expanded the physical plant, collection, and programs. The museum positioned itself for continued growth.

During her tenure, Dunbar was an influence on educational programs at several levels. As a flown astronaut she commanded respect, as a Ph.D. engineer she talked details, and as a woman she was particularly motivational for young women. Her friendship with Governor Christine Gregoire helped attract attention from the State and the Superintendent of Public Instruction. With many education programs in place or planned, she had the leverage to move them along. During this time, she also encouraged and supported Reba Gilman with the evolution of Aviation High School within the Highline School District. Although this was not an official project, the museum's support would prove crucial to the organization, land acquisition, and future funding for this incredible new kind of school.

Dunbar's focus was to align all of the 20 or so educational programs closely with national science, technology, engineering, and math (STEM) initiatives. This codified the museum programming in a direct and efficient manner and allocated resources to best serve objectives. She accomplished this organizational focus with staff and educational partners, building momentum. She also stressed the interpretation of technology and broadening of space themes.

Early in her tour, Dunbar, working with Governor Gregoire, launched Washington Aerospace Scholars (WAS) to promote STEM principles statewide.

Top: Museum of Flight CEO Bonnie Dunbar (right) teams up with Governor Christine Gregoire at a press conference on STEM education.

Bottom: Washington Aerospace Scholars with a 777 on a Boeing tour during their residency at the Museum of Flight.

Right: WAS students at the University of Washington's wind tunnel for part of their curriculum.

Above: Students inspect a model aircraft at Aerospace Camp.

This two-part program for high school juniors introduces them to NASA, space exploration, and the disciplines of earth and space science. The first phase, a five-month distance-learning course with content designed by NASA and the University of Washington, can earn a student five college credits. Phase two is a six-day residency at the MOF in which students apply what they have learned to plan a mission. It is an intense, competitive, and innovative program.

For younger students, Dunbar established the Aerospace Camp Experience (ACE). These full-day sessions, with some half-day sessions for younger children, host a mixture of aerospace activities, from model rocketry to building robots and even to a chance to fly. Accredited by the American Camp Association and an award recipient for its programs, ACE is tailored for all ages from preschool to ninth grade. These staff-intensive experiences squeeze a lot of fun and learning into a day visit and fill up early. In 2007, the program hosted a record 700 students in one week.

By 2006, programs touched 120,000 students around the state in alignment with STEM and the state Essential Academic Learning Requirements (EALRs). This outreach also served students in the formative Aviation High School. Foundations and individuals were interested in supporting learning programs, and the state, NASA, and others in government stepped up to fund curricula that met their objectives. The educational component of the museum proved it could attract broad support for capital to build and operate program funds. As courses ramped up, the MOF established a reputation for exceptional informal support to the formal school system. The museum's vision to create the "foremost educational air and space museum in the world" was the pacing commitment for the institution. It still is and will be into the future.

Budding astronauts plan their mission in the Space Station.

The international Association of Space Explorers (ASE) held their 21st Planetary Congress at the museum in September 2008, only the third time the assembly took place in the United States since its founding in 1985. 2008 also marked the 50th anniversary of NASA. ASE's mission is to encourage careers in engineering and science, with the hope that in the future, human life in space will be as customary as it is today on Earth. For the five-day conference, the museum's education department coordinated the visits of 50 astronauts and cosmonauts from 16 nations who traveled throughout Washington State to meet with 43,000 student and adult audiences to discuss space exploration. Public events at the museum included presentations on involving space sciences in the study of climate change and the environment.

Six leading informal science education institutions in Washington State united in 2009 to form the Washington Informal Science Education (WISE) Consortium, to advance pre-K-12 STEM education in the state. Members of the WISE Consortium are the Burke Museum of Natural History and Culture at the University of Washington, IslandWood on Bainbridge Island, the Museum of Flight, Pacific Science Center, Seattle Aquarium, and Woodland Park Zoo.

The consortium's diverse expertise is available to public school districts serving pre-K-12 students, universities, businesses, and other nonprofit organizations. The objective is to upgrade STEM literacy across the state, prepare students for scientific and technical careers, and meet the growing demand for workers in the tech industries, thereby enriching the state economy and ultimately helping a new generation to break out of a cycle of underemployment and poverty.

Education

Education is the compelling thread throughout the development of the Museum of Flight. Visitors are attracted to different aspects of the museum, but clearly, its educational power is its motivational force. Early in operations, the board made a formal commitment "To be the foremost educational air and space museum in the world." At this point, what was a longtime commitment was codified into the driving vision.

It is a long passage from a simulated airfield experience for young people at Seattle Center. The museum traveled from outreach programs with docents and discovery boxes to the Challenger Learning Center and formed a new level of experiential learning with the companion Aviation Learning Center, designed and built in-house. This base led to pathfinding STEM programs and finally to partnering with a formal Aviation High School. Education is the finest contribution of the Museum of Flight, and though its success is unparalleled, it is not a windfall, but rather an earned position based on early commitment, and promises made and not broken.

The focus of PNAHF founders was collecting and preserving regional flight history. It was that energy that initiated the museum organization. Once it was into a preservation obligation, education loomed in the planning, which would change the course of development.

Harl Brackin, a historian and industrial archivist, was committed to the dissemination of information, and early PNAHF planning documents featured public education. With the involvement of Jack Pierce of The Boeing Company, it was brought to the top of the priority list for enlisting corporate and community support for a flight museum. Bill Boeing Jr. talked the same language, underscoring this commitment. Founders signed on early and followed up with action.

The opening of the National Air and Space Museum in 1976 and its record-setting attendance proved the far-reaching popularity of aerospace themes. This new flagship energized the flight museum sector, and educators saw its potential. The MOF team was impressed. Here was an opportunity for a flight museum, with the high-quality interpretation of its themes, to inspire, stimulate learning, and motivate young people to pursue technical careers. From that point, there was no turning back. Education was ingrained into the museum planning, and this function that was initiated at Seattle Center infused all new museum development at Boeing Field.

The Museum of Flight, from its formative days, hired educators, trained docents to help with the process, reached out to schools, and found the budget to do all of this. It has been and still is a distinguishing element of the great success of the institution, but it was always a push.

It is a charter responsibility in Washington, and most states, to educate as justification for a nonprofit status. This function is expensive, and the costs cannot be entirely met by charging fees. These programs must be subsidized, and that can require organizational maturity. It is unusual for any museum to make these programs available from day one.

Over time, public programs and education gained attention, support, and funding, winning awards at regional, state, and federal levels. This service has grown in an accelerated fashion, breaking ground in its rapid evolution. The staff, intact through good and bad economic times, has established capability in curriculum planning and program implementation, often advancing leading-edge learning. The museum has developed both software and hardware. From the *Teaching Through Flight* resource manual, used by educators nationwide, to the Aerospace Camp Experience, the museum reaches hundreds of thousands of students annually.

Even with its focus on stimulating the young mind, the museum does not overlook service to visitors of all ages, the contribution to lifelong learning, and with its library and archives, the support to research.

Looking back to the initial education efforts, one would not think it possible to grow so much and so fast. The fact is that early supporters through to their contemporaries have carried the torch. It is fair to expect that this commitment will remain For Future Generations.

Students test an airfoil in the wind tunnel at the Aviation Learning Center.

During this stellar period in the expansion of the museum's aerospace educational programs, other areas of operations also experienced growth. The museum took on the challenge of building new facilities. As the land base expanded, planning ensued on the concepts for a space gallery and the long-awaited megastructure to house the collection of large aircraft.

Acquisitions continued on a regular basis. The MOF purchased a MiG-15 and MiG-17 in 2005. In 2007, Bill Boeing Jr. funded the replication of a Boeing Model 40B transport to grace the museum's commercial aviation story in the Great Gallery. The rare DC-2 rounded up by trustees Bruce McCaw, Joe Clark, and Clay Lacy back in 1999, and restored at Lacy's California operation, touched down at the museum with Captain Lacy at the controls in June 2007. This vintage aircraft would not be in the collection without the efforts of a vigilant acquisitions committee and a trio of longtime friends of the museum. The DC-2 is a remarkable find of a rare aircraft that many museums covet. This transport joins the Boeing 247 in revealing a pivotal phase in early modern commercial transport history.

Top: Addison Pemberton's Boeing Model 40C flies in formation with the museum's Douglas DC-2.

Bottom: The Lockheed Super G Constellation on arrival at Boeing Field.

The anxiously awaited Lockheed Super G Constellation made its debut at the museum in September 2009 after traveling overland from the headquarters of Empire Aero Corporation restoration facilities in Rome, New York. Believed to be one of only five commercial "Super G" aircraft in the world, the restoration was the result of more than four years' effort. Boeing executive and museum volunteer Bob Bogash took on this assignment to acquire what he considered "the most beautiful airplane ever built" for its "long, sinuous, curvaceous, and sensual" lines.

The indefatigable Bogash found this jewel abandoned after a second life as a restaurant in Toronto, Canada, an inglorious end for one of the beauties of commercial aviation history. He used his network of friends and contacts that ultimately led to support from the president of Air Canada. On behalf of the MOF, he was able to purchase the aircraft with the help of an anonymous donor. He then spent years pushing the project through the system, cutting

through Canadian government red tape for its release to the restoration company in New York. A letter to Bogash from the structures mechanic at Empire Aero concluded with "My dedicated crew and I have all embraced this project. Your airplane is in loving and capable hands at long last."

Arriving in the livery of Trans-Canada Airlines, the forerunner of Air Canada, the pieces were carefully reassembled. The magnificent "Connie" was put on display on the outside berm to greet museum visitors.

The P-51 Mustang completed its restoration in Idaho and returned to the museum in 2010, hailed by senior curator Dan Hagedorn as the finest work on this model of aircraft he had ever seen. The P-51 replaced a fiberglass replica that had been purchased to fill a void in the collection. Its perfect detailing and beauty fooled more than one former P-51 pilot. When the real deal was installed, many were eager to obtain the faithful replica.

The museum continued its presentation of cultural events that incorporated historical anniversaries, aerospace leaders, and valuable artifacts. These galas highlighted themes of military, commercial aviation, and space history, reaching out to build the museum image, attract supporters, and raise money for education.

On June 11, 2005, The Boeing Company presented the "Wings of Heroes" gala, celebrating the 35th anniversary of the Apollo 13 mission. Among those in attendance were astronauts James A. Lovell Jr., Fred W. Haise Jr., Thomas K. (Ken) Mattingly II, and mission control members, including flight director Eugene F. Kranz. With a surprise welcome and salute from astronaut Neil Armstrong, the evening was another success. William J. Rex, who contributed so much to the museum's galas over the years, chaired the event. Some 11 astronauts and a Russian cosmonaut participated, and the enthusiastic crowd raised their bidding paddles for $1.8 million for education programs.

Top left: Bob Bogash (left) shows off the "Connie" to Dominic Gates of the *Seattle Times*.

Top right: The Super G is moved across East Marginal Way at dawn to the Airpark.

Bottom: The fiberglass P-51 replica is removed, to be replaced with the real thing.

Top: Celebrating Apollo 13 (left to right): Master of Ceremonies Steve Pool, Jim Lovell, Fred Haise, Gene Kranz, Glynn Lunney, Milt Windler, Gerry Griffin, and T. K. Mattingly.

Bottom: The P-51 is introduced into the Personal Courage Wing.

The exhibits in the Personal Courage Wing, layered with interpretive materials, proved popular with visitors. Building on this success, the museum's internal design team, working at times with consultants, continued to enhance interpretive materials. In 2005, the museum unveiled new exhibits on both floors of the Red Barn with a focus on Boeing Company history. The company, so reluctant in the beginning to push for their involvement in the museum, was now prepared for and willing to support a much-expanded interpretation of their corporate history. Quietly, this was a breakthrough in museum-corporate relations. From this change, it was clear that the company recognized the value and quality of the new institution and had no reservations about being featured in its exhibits and programs.

The Boeing Story: 1916-1958 added interactive interpretation of the corporate story from its beginnings to the dawn of the jet age, with a full-size factory workshop illustrating the production of the Boeing Model C and Model 40. Remaining from the original exhibitry in the Red Barn when it opened for public display in 1983 was the office of early chief engineer Claire Egtvedt.

The in-house design group took on progressively larger tasks, adding value to the museum interpretations. The exhibit department worked closely with curatorial staff to create relevant displays and to produce high-quality presentations on modest budgets. Having a close connection to the artifacts and key stories, they also knew the daily exposure of the exhibits and could update materials economically.

While working on various options for land swaps and acquisitions to enlarge the museum campus at the start of the millennium, The Boeing Company offered another property on the west side that included Building 9-04. This stout structure was evaluated and found to be functional for housing the rapidly growing library and archives. In Autumn 2002, with modest remodeling, the department moved across the arterial roadway and established a beachhead on the west campus. In that location, it had the space to grow, which it did, with new collections added to remodeled facilities in the structure.

The Kenneth H. Dahlberg Military Aviation Research Center opened as part of the archives in 2006, named in honor of the famous fighter ace and entrepreneur. With the enlarged facilities, the library and archives progressed at a record pace in receiving and cataloging donations, including those of the American Fighter Aces, building up one of the best aerospace history centers in the world. This extensive collection was organized to be more accessible, with manual labor aided by the application of advanced management software. What was available to serious researchers on site was now accessible online via the Internet.

With the earlier Wright brothers traveling exhibit, the museum had found it could successfully host these attractions, adding custom touches. In some cases, these were the best kind of currency for the museum. A special exhibit on the genius of Leonardo was seized upon as one such opportunity.

The *Leonardo da Vinci: Man, Inventor, Genius* traveling exhibit had its West Coast premiere in October 2006. Celebrating this Renaissance genius, the interactive exhibits brought in new audiences and added revenues while building the image of the institution. The show featured 50 3-D models of the machines that da Vinci designed, along with dramatic copies of his artworks. Bill Gates, the founder of Microsoft, loaned a page from da Vinci's Codex Leicester. Securely displayed, this incredible document added authenticity to the exhibit.

Playing the part of Leonardo da Vinci, museum educator Richard Wallace interprets a model for Geda and Phil Condit during the grand opening gala for the da Vinci exhibit.

EMS Entertainment, headquartered in Vienna, Austria, organized the traveling da Vinci exhibition. Board member McCaw met with the CEO of that company at the MOF to discuss terms. McCaw was captivated by the beauty of the da Vinci artwork in the portfolio. Soon after, traveling to Vienna on other business, he had coffee with the CEO and discovered that these reproductions were with a companion exhibit focusing on da Vinci's artworks. This *Man, Artist, Genius* display featured digital reproductions of 23 da Vinci paintings, including "The Last Supper" and "Mona Lisa," replicated in the original sizes to a high degree of authenticity. McCaw inquired about the possibility of their inclusion in the *Inventor* exhibit scheduled for the museum. He made the deal, and the reproductions were shipped to Seattle. This blending of the art with the display of replicated inventions added yet another dimension to the exhibition.

The museum staff is adept at layering interpretive materials to dramatize displays, and their creative work enhanced the exhibit's appeal. They used a timeline chronicle to tie together the interactive displays of da Vinci's inventions for flight, mechanics, hydraulics, and weapons. The magnificent painting reproductions were placed separately and hung carefully for lighting control and dramatic effect. The combination of art and invention worked well in the flight museum context.

Educational programming for the exhibit featured curriculum materials that supported the Essential Academic Learning Requirements (EALR) of Washington State. These were distributed electronically, allowing for pre- and post-visit activities.

The Simonyi Fund for Arts and Sciences sponsored the opening gala for Leonardo, which raised another $1.4 million for education. Three members of the Seattle Opera performed Italian arias for the more than 400 attendees. This exhibit reached out to a broad audience within the cultural community. The museum continued to mature as an aerospace center that touched upon art, science, history, and culture, creating more opportunities for the educational mission.

The evidence of the in-house exhibit talent was also on display in the impressive dioramas and multimedia interpretations developed in 2007 for *Space: Exploring the New Frontier*. The exhibition showcased more of the museum's space artifacts, with interactive experiences such as landing a space shuttle or lunar lander.

In 2007, Chair Robert Genise reported that the museum's net worth had swelled to $100 million. Riding a strong economy, this was a solid year all around, with a surge in visitation and earned income, a banner membership of 23,000, and a growing number of students in more than 20 formal aerospace programs. The museum was named an Affiliate of the Smithsonian Institution, a partnership that affords better access to the national collection to qualifying institutions.

The euphoria of a banner year was not to last long. In 2008, the country was experiencing the worst economic recession since World War II. The downturn was not unexpected, but it lasted longer than the experts predicted. The result was a drop in the stock market, a credit freeze, employee layoffs, and the failure of financial institutions. The effect was global. Seattle, characteristically slower to feel the damage, was hit hard. A region that had experienced the boom-and-bust cycles of aerospace was now in a punishing recession. The impact was more damaging than the Nisqually earthquake of 2001. Nonprofits were battered during this period in which wealth and philanthropic contributions receded along with the dip in earned income.

The Museum of Flight faced severe budget reductions that year, and for several years to come, and adjusted accordingly. The biggest budget item, staff salaries, was identified for a 10% reduction. Difficult as that was, the museum moved quickly with cutbacks in size and compensation, a critical move that employees

A crowd watches as construction cranes carefully set the main span of the bridge into place.

volunteered to support. Even with budget cuts, the museum did not intend to sit back and await better times. It could not delay obligations for program enhancement and physical expansion. Dunbar inherited this challenging period, and with belt-tightening on staff and support of the board, the museum found ways to continue crucial commitments.

The sound base set with the opening of the Personal Courage Wing helped to sustain programming and operations over this tough patch. Chair Genise reported that the challenging year was weathered comparatively well. The good news in these bad times was that the museum could adjust and find ways to perform.

With the land acquisition and the expansion of the campus both north and west, the museum began to plan the next phase. The growing pedestrian traffic crossing the busy arterial was a concern, and all recognized the need to provide better access and a protected passage from the east to west campus, to the archives and large-aircraft storage. It was time to address that issue while also integrating the museum campus.

The T. Evans Wyckoff Memorial Bridge connects the East and West Campuses.

Fortunately, government funding sources were available because of the safety issue. U.S. Senator Patty Murray from Washington State worked on a $2 million appropriation in a transportation bill; this was matched by the State, with another $500,000 from King County. This was a good start on funding for the $10 million bridge, with the rest of the money contributed privately.

As funds were put together in a fast-paced campaign, a board member stepped up unexpectedly with a $5 million gift in memory of his friend T. Evans Wyckoff, joining additional contributions from the Wyckoff family. Wyckoff had been an early trustee on the museum board, and it was appropriate to affix his name to this essential and beautiful connector. Board members Ed Renouard and Gene McBrayer, nicknamed the "Tag Team" for their proven ability to get things done, worked on this project, and it came in on budget and schedule.

While it was not the most expensive or largest expansion, the T. Evans Wyckoff Memorial Bridge was a critical component. This unique structure is more than just an elevated arterial crossing; it is a unique and dramatic experience in sight and sound that beautifully connects the enlarged west campus with the airfield facilities.

The award-winning SRG design treats the bridge as sculpture, simulating a swirling vortex. Artist and musician Paul Rucker created the sounds and "Trails of Vapor" music that resonate from speakers along the bridge. It is a functional structure that provides an unusual experience for the user. Ed Renouard said that the cocktail party on the bridge for its grand opening in 2008 was one of the best in his memory.

Tag Team

The best of plans are static until executed. Implementation has evolved into one of the strengths at the Museum of Flight. The so-called "Tag Team" of board members Ed Renouard and Gene McBrayer personifies the committee structure that efficiently ties planning to doing. McBrayer joined the board in 1997 and Renouard in 1998. At the time they had not met, but would soon connect and with their considerable and complementary skills forge an efficient process for building.

Ed Renouard is a retired Boeing executive who managed both the Auburn facility and the huge Everett plant. In his career, he gained experience in the construction of large facilities. In 1998, Phil Condit, CEO of The Boeing Company, asked him to help the museum with its expansion. He joined the board for that task and then became more active as there was a growing need for his expertise. At the invitation of Gene McBrayer, Renouard acceded to the building committee to evaluate a proposed master plan. That is how he started, and what he still does.

The other member of the Tag Team is Gene McBrayer, a graduate in chemical engineering who worked his way up to become president of Exxon Chemical division. In his career, McBrayer also had experience in building plants at several locations. He retired early and moved to Puget Sound, as he and his wife had always liked the area. McBrayer pursued his passion for flying, meeting a museum trustee, Mark Kirchner, chair of the MOF, who asked McBrayer to join the board.

With his business perspective, McBrayer served initially on the finance committee. In 2000, chair Harold Carr asked him to take over a building committee, with Dick Taylor still in charge of planning. This move initiated the governing board's larger role in the project management of new development.

McBrayer's new role came as the board faced up to a proposed $140 million master plan. Their concern about the size and cost of this expansion was made worse by the economic downturn of dot-coms and then the 9/11 issues. McBrayer welcomed the challenge of putting together a process that would identify viable projects, attract funding, and then use those resources wisely.

From his experience, McBrayer knew that every region of the country had its own way of building. He asked friends about the best method to do the job in Seattle and was referred to a principal at Seneca Group. This development and management company had steered several large projects to completion within budget and schedule, also rescuing some that were off course. His discussions with managers at Seneca resulted in the establishment of a design-build team that could introduce efficiencies and quality engineering.

McBrayer organized the team, with Seneca Group as project management counsel, NBBJ as architects, and Sellen Construction as the general contractor. When the museum project architects shifted from NBBJ to SRG, the new firm was selected for the team. All of the principals have now worked together for years on many projects.

Building and planning began to whittle away at the whole master plan and divide it into stages to create more manageable tasks for funding. Each piece was also evaluated for priority of need and organized into viable phases. The process became a series of practical steps. The big test for this process was the proposed $140 million expansion.

The Tag Team and consultants reviewed the development against funding potential and priority needs and found it financially not feasible at the time. The team reduced it to a four-phase plan, the first of which would change several times due to land deals and emerging opportunities. This strategy allowed the expansion to begin and establish momentum in funding and building. The process is flexible and agile, adjusting to changes in requirements. Each successive phase takes the same track, the evaluation based on cost and priority, with each also experiencing course corrections. The implementation strategy is coupled with the experienced design-build team approach to saving time and money.

This process has worked to keep the capital projects in the pipeline and construct needed facilities. The alignment of planning and building has worked effectively, bringing in skills from other committees as necessary. The result is efficiency and progress and returns on investment. Those who contribute feel comfortable that money is well spent.

The Tag Team has helped all of the expansion projects for more than 15 years: the Personal Courage Wing and associated improvements; Wyckoff Bridge; Simonyi Space Gallery; and the Aircraft Pavilion. They worked up close on these additions that total more than $92 million in development. Their leadership energizes others. If one of the team is fatigued, the other comes over the ropes to take over. These volunteers are a force.

The foundation is eager to point out that this hands-on expertise pays off, with all of these projects on target, most coming in ahead of schedule and under budget. It is hard to put a value on this resource, but it is an organizational strength that is the envy of any nonprofit.

The "Tag Team," Ed Renouard (right) and Gene McBrayer (center), with Wilf Wainhouse of Sellen Construction celebrating the Aviation Pavilion completion, 2016.

In 2009, new chair J. Kevin Callaghan reported on how the museum had survived and even thrived in a tough economy, now leaner but still healthy and delivering service. Trimming some $3 million from the budget offset a 20% decrease in earned income to operations. Even with a drop in program participants, the numbers posted were substantial in service to the community. The museum cruised ahead through this trying period.

After reevaluation, AAM reaccredited the museum, maintaining and assuring its integrity. The MOF also joined the Association of Science – Technology Centers (ASTC), an organization currently representing more than 600 institutions in nearly 50 countries.

The museum was in full rebound by 2010, "a great leap forward, overcoming challenges of recession," as chair Callaghan put it. Attendance was up again, best since the slump, and earned income was flowing. The remaining debt for the Personal Courage Wing, a construction loan, was paid back in full with another crucial and generous donation from Bill Boeing Jr.

June 12, 2010, honored the museum's magnificently restored B-17F upon the 75th anniversary of its first flight. The B-17 was targeted early as a must-have centerpiece because of all its power and strength provided to the war, to the Seattle economy, and to aviation.

The "Boeing Bee" is the centerpiece at the 2010 Hangar Dance to celebrate the 75th anniversary of the B-17's first flight.

Although initially restored for flight before delivery to the museum, the aircraft had a lot of deficiencies. The BMA volunteer restoration crew included a team obsessed with authenticity and high quality. For more than a decade, the group addressed every detail inside and out. The resulting object is a piece of art and one of the finest B-17 examples on Earth.

The B-17F *Boeing Bee* was the star of the evening. The program referred to her as "A hometown hero that was the last of her kind to leave the famed Boeing Plant II. She's been an aerial sprayer, firefighter, tanker and war memorial." Now she was in the spotlight and ready for her new museum home, and she was in the right company with John Sessions providing a fly-by salute at the controls of his rare B-25D in formation with two P-51D little friends from collections at Paine Field. It was time to celebrate.

What a party it was. Clay Lacy Aviation and Aviation Partners sponsored the bash, held at Lacy Aviation hangar and chaired by Nancy and Charlie Hogan. The period motif was vintage World War II. Gala regular Steve Pool served as emcee, with Fred Radke leading and Gina Funes singing with the Harry James Orchestra. As if one band was not enough for the evening, Bill Tole directed his Jimmy Dorsey Orchestra as the period-dressed crowd took to the dance floor. The

Pied Piper Singers and the Swing Dolls, covering Andrews Sisters tunes, kept up the beat.

On the serious side, the occasion saluted all who served with this legendary warbird, "honoring the men and women who designed, built, ferried, protected, and flew her" as well as the team that spent thousands of hours restoring the artifact. The crowd extended a long standing ovation as representatives of all these groups, including the Tuskegee Airmen and Women Airforce Service Pilots (WASP), took the stage.

The evening contained a special tribute to Dick Friel, who passed away in January 2010. Friel, the celebrity auctioneer, noted emcee, aviation pioneer, and MOF supporter, and his wife, Sharon, had together helped raise hundreds of millions for charities during his distinguished career. Sharon took over the auctioneer duties for the evening.

This gala exemplified the events that have evolved at the museum. It had a precious artifact as a centerpiece, surrounded by historic aircraft and period décor. It included people who served with the B-17 and paid tribute to its stalwart performance. The event was commemorative, it was fun, and as a result raised lots of money.

The museum had taken a quantum leap from initial operations and was enhancing the space collection, adding exhibits, attracting astronauts and pioneers on a regular basis, and creating galas of commemoration. No institution in the country was doing a better job of featuring space stories and educating young people on the subject. The museum was meeting its objectives in reaching out to more than 100,000 students annually, promoting STEM learning and motivating young people to pursue careers in aerospace.

It was clear that the MOF was forging new paths in space themes and interpretation. What could be a better case in asking for respect from NASM in the form of a Space Shuttle? One was soon to become available.

Top: Attendees dine in the Clay Lacy Aviation hangar with the Flying Fortress as a backdrop.

Bottom: PNAHF founding trustee H. C. "Kit" Carson (left) is recognized along with other B-17 veterans at the gala.

Bonnie J. Dunbar, Ph.D.

President and CEO, the Museum of Flight, 2005-2010

I knew that I wanted to be a part of the Museum of Flight, its commitment to excellence, and its extended and outstanding team when I was invited in 1987 to speak during the opening of the Great Gallery by Boeing Chairman T. Wilson and MOF Executive Director Howard Lovering. My host that day was Boeing VP Dick Taylor, a pioneer in his own right, who in the following years helped me to appreciate the amazing journey that aviation has traveled in less than a century.

President and CEO Bonnie J. Dunbar in the Great Gallery.

Although I was well versed in the history and technology of space flight, and had a private pilot's license, the pioneering stories of early aviation in Washington State and the evolution of The Boeing Company were not as well known to me, a ranch girl from a small community in Eastern Washington. However, in the following years, whenever there was any invitation forthcoming to the Astronaut Office to participate in MOF activities, I heartily volunteered. A few years later, I was very honored to be invited to join the board of trustees. Although I was not always able to attend board meetings from Houston, Texas, I followed the MOF closely.

In early 2005, I was invited by the board chair to participate in the search for a new president and CEO with the retirement of Ralph Bufano. A few weeks later, I received another call, asking me if I would withdraw from the search committee and add my name to those under consideration. At the time, I had completed my Shuttle flying career and was in the Government's Senior Executive Service, had recently completed the Harvard Senior Managers in Government program, had occupied several NASA management positions, and was on a trajectory to a full executive career with NASA.

Retiring from NASA was not on my radar, but with this opportunity I saw a different path that would merge my love of flight with education and a return to the Northwest. I was not optimistic about my selection, but I couldn't think of a better opportunity to try. During the interviews with the board and with staff, I was pleased to learn that the next horizon for the museum was to include more youth education (later part of the national STEM initiative) and to grow our space-exploration history, collections, and exhibit presence. I couldn't have been more pleased when I received the phone call that I had been selected.

During the next five years, our board and donors committed themselves to growing our educational offerings, consistent with the federal government's leadership in science, technology, engineering, and mathematics (STEM) education in the Informal Science Education (ISE) community. As a member of the National Academy of Engineering (NAE) education committee, I was able to reinforce the value of having museums and science centers recognized as contributors to the national solution of inspiring our youth to pursue engineering, in order to keep us economically competitive and strategically safe.

Early in 2006, I assumed Ralph's position on the board of the Aviation High School (AHS), where I met an engine of education, Reba Gilman, the principal of this new school. AHS was and is a public high school in the Highline School District, open to any student in the state of Washington, with a maximum enrollment of 400 and a focus on *aerospace* as inspiration. However, AHS was in a temporary space on the grounds of a community college, which in 2006 was ready to reclaim its buildings. When it became apparent that AHS might disappear, the board of trustees of the MOF stepped forward, and with the help of trustees Bruce McCaw and James Raisbeck, AHS was eventually built adjacent to the Museum of Flight. In 2016, it was ranked by *U.S. News & World Report* as the #1 Public High School in the state of Washington, with nearly a 100% graduation rate.

In parallel, we migrated a statewide high school program from NASA Johnson Space Center (NASA JSC), called Texas Aerospace Scholars, to Washington State, with the assistance of trustee Bill Boeing and the governor of Washington, Christine Gregoire. Rebranded Washington Aerospace Scholars, or WAS, the program has steadily grown in popularity, support, and STEM recruitment. A major milestone was reached when the University of Washington awarded incoming freshmen who had attended WAS an automatic three credits toward graduation.

"Space" remained on our radar. With the help of a generous gift from trustee Bill Boeing and an inspired director of exhibits, Chris Mailander, we repurposed the side gallery into a permanent exhibit on the history of space flight, utilizing every artifact we had collected and bringing in a few more, such as the high-fidelity ISS Research Module "Destiny" and the engineering test article/backup for the Mariner Mars lander. While we were not successful in our bid to land a retiring Space Shuttle, we did build a new Simonyi Space Gallery, housing the Full Fuselage Trainer (FFT), which trained every astronaut who ever flew on the 135 Space Shuttle flights.

Our team achieved much in five years, building on the prior achievements of the board of trustees, Presidents Lovering and Bufano, and a devoted and talented staff, volunteers, docents, members, and donors. We grew our commitment to educating youth, to engaging the community in new ways, including social media, and to enlarging our space collection.

However, this period was not without its trials and tribulations—most notably how we managed to shoulder our way through the 2008 recession, when many nonprofits did not survive. We were able to succeed because of the leadership of our chairman, who advised me early on to hold tight on personnel levels and expenses based on what he could see in the global financial markets; the commitment of the staff, who remained focused on excellence and on a quality visitor experience but often had to endure flat compensation levels; and a supportive board. Along the way, we also restored a Lockheed Constellation and a P-51; grew our helicopter and glider collection; paid tribute to our veterans from WWII, Vietnam, and up to the present; dedicated an Alaska Airlines Exhibit; and refurbished our Tower/Aviation exhibit. I have moved on to other ventures, but my five years at the Museum of Flight have remained some of the most rewarding of my life.

— Bonnie J. Dunbar, Ph.D.

Michael and Mary Kay Hallman Spaceflight Academy in the Museum of Flight's Charles Simonyi Space Gallery.

CHAPTER SEVENTEEN

Rush for a Shuttle

At the end of 2009, the museum found itself in competition with 20 other museums and science centers for one of the three retired NASA Space Shuttles. The dream of a Shuttle to grace the rapidly growing space collection was in sight. No one associated with the museum thought twice about focusing on this competition and winning. Everyone agreed the game was on.

An organizational structure was put in place to go after a Shuttle as an appropriate centerpiece to everything the museum was doing in aerospace education. Dr. Bonnie Dunbar stepped down from CEO duties to lead the competition as well as the emerging partnership with another project, Aviation High School, and other STEM initiatives.

Dunbar had brought to the museum unique expertise as a flown astronaut, engineer, and nationally acclaimed educator focused on these priorities. She immediately transitioned to CEO of the Wings Over Washington organization, founded back in 1989 at the time of the state centennial and now in place to support specialized assignments for the museum. These duties would need full attention, and there was no taking chances.

Always ready to help, trustee Mike Hallman accepted the job of museum interim CEO. He was a comfortable fit, with his foundation knowledge and long experience as an executive. During this transition, the museum didn't miss a step in continuing its public service.

A keystone in NASA's competition for the reward of a Shuttle was to meet the requirement of a climate-controlled gallery for its display, and museum leadership thought it wise to step up and build the space pavilion, no matter what the outcome. This was a gutsy decision, but the right thing to do.

The MOF did have a prior commitment for the Full Fuselage Trainer (FFT) component from NASA, which was a fallback though not a prime conversation in the team's full-out proposal. The museum team envisioned the Space Shuttle as a companion to the FFT in what would be one of the leading space interpretations in the country. It was a grand vision so typical of the museum leadership.

Charles Simonyi

On April 8, 2009, Charles Simonyi returned to Earth from the International Space Station. It was actually the computing pioneer-cum-philanthropist's second round of space tourism there, the first occurring in 2007. Both spaceflights perfectly capture the irrepressible spirit of a brilliant technical innovator who takes great joy in life and pleasure in making the world a better place.

Born in Budapest, Hungary, in 1948, Simonyi found his life's professional focus during his high school years, when his father, a professor of electrical engineering, arranged for him to assist at a computing facility. It was the 1960s, when mainframe computers were enormous and programming was primitive. The boy soaked it all up and learned quickly under the kind tutelage of a computer lab engineer with a genius for explaining the elegant mathematical underpinnings of computing.

Simonyi loved it. Using the Soviet-built Ural II computer gave him tactile pleasures even as it posed challenges for him to think and work his way through. By the time he graduated from high school, he had already written his first compiler, which translated high-level programming into simpler code that served as an executable program. This success led him to Denmark at age 17, where he gained initial experience in his chosen profession while saving up for further education.

California beckoned as the global nexus of computing progress. Arriving in the United States at age 19, Simonyi enrolled at the University of California, Berkeley, and graduated in 1972 with a bachelor of science degree in engineering mathematics and statistics. Not yet 24, he found himself in demand at a time when computing and related technologies were advancing exponentially. While continuing his studies, he worked at the UC Berkeley Computer Center, Berkeley Computer Corporation, the ILLIAC IV Project for parallel computing, and Xerox's formative Palo Alto Research Center (PARC).

Simonyi received a doctorate in computer science from Stanford in 1977 while at the same time helping Xerox PARC develop the Xerox Alto, one of the first personal computers. In 1981, he accepted Bill Gates' invitation to join Microsoft and crate its applications group. Simonyi gathered and led the team that invented Microsoft Word and Excel, which are perennially the most popular applications in the Microsoft personal and business computing software suite.

After two decades with Microsoft, Simonyi departed in 2002 to cofound Intentional Software, which markets the intentional programming concepts he pioneered at Microsoft Research. A talented innovator who holds many patents, he is globally recognized for his profound influence on programming and information technology. He currently chairs the board of trustees of the Institute for Advanced Study. In his spare time, Simonyi enjoys flying as a licensed helicopter pilot.

In 2008, Simonyi married Lisa Persdotter in a private ceremony in Gothenburg, Sweden. Today, they are the proud and happy parents of two beautiful young daughters. Continuing Simonyi's longtime philanthropy, they also oversee the Charles and Lisa Simonyi Fund for Arts and Sciences, which provides grants in arts, science, and education to organizations in the Seattle region and around the world. The Seattle Symphony, Seattle Public Library, Woodland Park Zoo, and Seattle Art Museum are among the beneficiaries of the Simonyis' dedication to giving back to the community in ways that promote access to excellence.

At the Museum of Flight, this philanthropic support was instrumental in creating the state-of-the-art Charles Simonyi Space Gallery, which bears his name in recognition of his commitment to aerospace education and enthusiasm for inspiring the next generation of space explorers. Fittingly, one of the artifacts displayed there is the Russian Soyuz TMA-14 spacecraft that carried Simonyi to the ISS in 2009.

Charles Simonyi and his wife, Lisa, at the dedication of the Charles Simonyi Space Gallery in December 2011.

An aerial photo of the west campus (lower half) with room for expansion, 2006.

The museum chose the west campus as the best place to construct the new space gallery. The price tag of $11 million was met fairly quickly with state money, a contribution of $1 million from the planning chair, another $1 million from the board chair, plus other donations, rounded off with a major gift from the Charles and Lisa Simonyi Fund for Arts and Sciences that finished the pledges for the full amount. With the private funding assembled in an accelerated campaign, the project attracted a State Department of Commerce grant as well as U.S. Treasury New Markets Tax Credit financing. With this gallery defined as an exhibit, the project also qualified for state sales tax relief that helped in the funding equation and saved a lot of money.

The museum announced that the Space Gallery would break ground on June 29, 2010, a fast track from the decision to design and funding. From start to finish, the 15,000-square-foot gallery took 16 months to complete and was tailored to accommodate a flown space shuttle and the FFT. The consultant team of SRG Architects, Sellen Construction, and Seneca Group went into action with the building committee and completed the project on budget and schedule.

The competition process made it clear that the Smithsonian National Air and Space Museum, which already displayed the *Enterprise*, would keep it or get priority for a trade. Also, simple logic suggested that one of the Space Center attractions would have collegial points from NASA, with the *Atlantis* going to the Kennedy Space Center Visitor Complex in Florida. But that left two slots up for grabs. As the MOF was the foremost educational air and space museum in the world, its proponents expected to have an advantage.

The request for information (RFI) and guidelines mentioned that the competition would be objective, not political, and that one requirement was a suitable structure in which the shuttle could inspire and educate. No other competitor was building a space gallery, so that increased the likelihood of the MOF's selection. The museum team put together a detailed proposal focusing on institutional strengths regarding all criteria of the competition and underscoring the new Space Gallery that was in development.

The formal announcement of the Shuttle awards in April 2011 held surprises. Udvar-Hazy Center, the NASM annex, received their desired shuttle, *Discovery*, as was expected, moving their current display of the *Enterprise* to the Intrepid Sea, Air, & Space Museum in New York. The *Atlantis* was awarded to the Kennedy Space Center Visitor Complex, near Cape Canaveral, Florida, also as anticipated. The last and final, *Endeavour*, was not on its way to Seattle, but to the California Science Center in Los Angeles.

In the announcement, NASA pointed out that additional artifacts were distributed to several competitors. The Full Fuselage Trainer was destined for the Museum of Flight. The decision was down.

Chair Callaghan reported, "While we weren't selected to be the home of one of the retiring Space Shuttles, we have already put together plans for an extraordinary exhibit that will further the educational mission of the museum, telling the story of not only the past of space flight, but the future as well."

Public announcements were civil and measured, touting the Full Fuselage Trainer as a valuable artifact, one that could do a signal job of interpreting astronaut training for shuttle missions, and one that the public could access. All of this was true, but there was still a lingering discontent about the decision.

The chair of the NASA committee that made the selections pointed now to the guidance from Congress that the orbiters go to facilities where the most people could see them, though this was not stressed in the initial request. The selection chair justified the California decision based on ties to the space program of Southern California and Edwards Air Force Base, as well as plants that manufactured components. The Smithsonian, of course, was the official curator of

Top: Artist's concept of the Charles Simonyi Space Gallery.

Bottom: The Space Gallery begins construction and anchors the west campus next to the Airpark.

space artifacts. The Kennedy Space Center Visitor Complex was where launches originated, and Intrepid Sea, Air, & Space Museum had served as a recovery ship for Mercury and Gemini missions. All of these were also rated highly for exposure to population base and tourism. It appeared that the rules were clarified after the selection, as there was little if any mention of aerospace education.

If NASA was committed to educating the public about its mission, how could the agency overlook the leadership demonstrated by the Museum of Flight? Over the decades the MOF had embraced NASA initiatives and space education, already reaching hundreds of thousands of students. As one board leader mentioned, the FFT is not an option or consolation, as the MOF should have both in the Space Gallery, the trainer adding interpretive value to the flown Shuttle in a way that would capture the imagination of visitors to the new Gallery and provide constant inspiration to students in the programs. Ultimately, its value was not just as a popular attraction, though it certainly was destined to be just that, but also as a teaching tool for future generations.

Space Shuttle FFT

Full Fuselage Trainer in NASA's Building 9NW at Johnson Space Center before being moved to the Museum of Flight.

The NASA Space Shuttle Full Fuselage Trainer (FFT) provides a crowd-pleasing centerpiece for the Museum of Flight's Charles Simonyi Space Gallery. This huge one-of-a-kind training device delights visitors by providing access and understanding that would not be possible with an actual Space Shuttle.

Although the Shuttles went elsewhere, NASA separately decided that the Museum of Flight would receive the FFT. This huge wingless mockup—a prized artifact in its own right with a long and rich history—measures 122 feet (37.2 m) in length and more than 46 feet (14 m) in height. It represents in full scale and proper relationship all the parts of a Space Shuttle that human beings might live or work in during spaceflight.

Over several decades, the FFT familiarized astronauts with the Space Shuttle's systems and afforded them realistic training in emergency egress, extravehicular activity (EVA), and other procedures. Every astronaut who flew aboard a Space Shuttle spent countless hours training in the FFT and knew it intimately. Although this NASA training device incorporates some lighting, payload bay, and other systems that work, its flight deck and most other systems do not, since it is a procedures trainer and not a flight simulator.

The FFT's previous home was the NASA Space Vehicle Mockup Facility (SVMF) at Johnson Space Center in Houston, Texas. Astronaut training at the SVMF typically took at least a year and sometimes longer, depending on the objectives of the mission. Each crew spent up to 100 hours in more than 20 separate classes. Concurrently, the FFT also served as a valuable testbed for NASA engineers and specialists, who used it to quickly and easily evaluate and refine in situ a host of upgrades to the in-service Space Shuttle fleet.

The FFT plays an important role in the 15,500-square-foot (1,440 m²) Charles Simonyi Space Gallery, which combines state-of-the-art interpretation with hands-on learning. Visitors can enter the Shuttle trainer's payload bay and, in special cases, tour its mid deck and flight deck accompanied by a knowledgeable docent. The result is a high degree of interpretive potential that promotes two key Museum of Flight goals, these being to educate the public and inspire the young to pursue STEM careers. In this sense, the FFT is today still at work turning out future astronauts.

Only at the Museum of Flight can visitors get a pilot's perspective of the Space Shuttle.

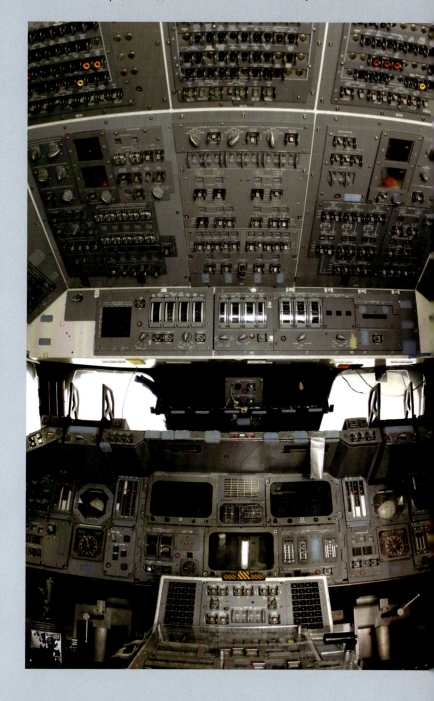

The museum leadership asked for and was granted a debriefing in Washington, D.C., which amounted to the selection team reiterating what had been broadcast about the decision—that it was a hard call but fair, and it was over and done. In the words of the NASA administrator: "This was a very difficult decision, but one that was made with the American public in mind. In the end, these choices provide the greatest number of people with the best opportunity to share in the history and accomplishments of NASA's remarkable Space Shuttle Program. These facilities we've chosen have a noteworthy legacy of preserving space artifacts and providing outstanding access to U.S. and international visitors."

Other competitors were disappointed enough to call for congressional insight into the process. In particular, the Johnson Space Center in Houston and National Museum of the United States Air Force called for another look at the competition, which was not to be. Some of the institutions could make strong cases based on individual strengths. The MOF did not enter this fray, though it had a long list of particulars with which to make a case.

The fact also remained that the Museum of Flight was the only competitor that met the NASA requirement to have a display building in place for the award. Ed Renouard recalled that he felt "righteous indignation," explaining that Dr. Dunbar's team had completed the checklist, built a facility, offered the best space education programs, and submitted a good proposal. He did find consolation in the fact that the FFT is one-of-a-kind, exciting to visitors of all ages, and that the Simonyi Space Gallery was such a striking addition to the museum complex.

The museum team had completed a challenging task and felt the crush of disappointment. But with so much going on, this emotion was not allowed to slow the work under way. Museum officials put a positive spin on the occasion and went about completing the Space Gallery, which would soon be full of incredible artifacts, old and new, with exhibits and programs that are the envy of the museum world. The museum did not lose the competition; it just did not get a

Charles Simonyi addresses guests at the dedication of the Charles Simonyi Space Gallery in 2011.

Shuttle. It was the retired Shuttle that lost a perfect place to reside.

On December 8, 2011, the Charles Simonyi Space Gallery celebrated its grand opening on the west side of the museum campus. The gallery honored the two-time space traveler and software pioneer. Dr. Simonyi and wife Lisa Persdotter Simonyi joined the dedication along with elected officials and dignitaries.

Simonyi generously donated $3 million to the Space Gallery and the Soyuz TMA-14 descent module from his second space flight in 2009.

In presenting the capsule, Simonyi explained, "This was the capsule that brought me, along with two crewmates, back to Earth from the International Space Station. It sustained and protected us as we reentered the Earth's atmosphere and landed on a prairie in Kazakhstan." Along with the capsule, Simonyi also loaned his spacesuit and a space toilet.

Also displayed in the pavilion is the Charon, Blue Origin's first vertical takeoff and landing test vehicle. On loan from Blue Origin, the 9,500-pound Charon used four jet-powered engines on its test flights in 2005. The facility is the new home for the museum's Apollo space capsule and Martian lander, and its interactive, multisensory exhibits tell the stories of all the visionaries, designers, pilots, and crews involved in the space program.

Chair Hallman and a new museum CEO, Doug King, reported a "giant leap" with the dedication of the Charles Simonyi Space Gallery and referred to it as an addition that is an appropriate kickoff for the next 50 years. They wrote, "Today as we await delivery of the NASA Space Shuttle Trainer, we look forward to realizing the bigger picture. The completion of exhibits in the Charles Simonyi Space Gallery will signal the beginning of a new era at the Museum of Flight. There, we will tell the stories of the past 50 years of

space travel and the next 50 as well. We will share the aspirations of adventurers, innovators, and pioneers who took the first steps off our planet . . . and we will look to inspire the generations to come that will walk in those individuals' footsteps."

Museum representatives traveled to Johnson Space Center in January 2012 for the formal transfer of the Full Fuselage Trainer. When the crew compartment flew in on a delivery flight in NASA's "Super Guppy" transport on June 30, 2012, the museum celebrated "ShuttleFest." Governor Christine Gregoire welcomed the FFT along with former Governors Mike Lowry and Dan Evans, Congressman Jim McDermott, five astronauts, and George Mueller, the "Father of the Space Shuttle." Board member Clay Lacy, who had test flown the very first pregnant Guppy transport, was also on the tarmac. The Seattle Sounders band played for the cheering crowd.

For the last 30 years, the wooden and wingless FFT had trained every U.S. space shuttle astronaut, and soon it would prove instrumental in accelerating the museum's space education programs into the future. The museum now had a beautiful new gallery, an accessible and authentic trainer, and the beginning of a space interpretive center.

Top: (left to right) Johnson Space Center Mission Operations Director Paul Hill and NASA representative Greg C. Johnson hand over the "keys" of the FFT to Museum of Flight CEO Doug King.

Bottom: The FFT nose section being removed from NASA's "Super Guppy" at Boeing Field.

Michael R. Hallman
Interim President and CEO, the Museum of Flight, 2010

My association with the Museum of Flight began in 1990, when I was asked to be a trustee. At that time the museum consisted of the Great Gallery and the Red Barn and a terrific collection of airplanes that appealed to a large number of aviation enthusiasts in the Puget Sound region. I have been incredibly fortunate to have a front-row seat to watch this museum grow and evolve into the "foremost education air and space museum in the world." Over that 26 years, I chaired the Exhibit Committee for 10 years, and served as Interim CEO in 2010 as well as Vice Chair and Chairman of the Board of Trustees. Looking back, I am amazed at how far and how fast we have come.

I have always had an intense interest in education, but my passion for aviation was different from that of my colleagues on the board. Rather than being a pilot, engineer, or in airline operations, my background was in computers and software, having worked in sales and marketing for IBM for 20 years. My aviation interests grew when I joined Boeing as CIO just as they began to develop the 777, using CAD (computer-aided design) software more intensively than ever before.

I tend to look at the museum from the perspective of the main inflection points toward our strategic goals rather than just the acquisition of specific artifacts.

The Personal Courage Wing, opening on the 60th anniversary of D-Day on June 6, 2004, was a significant step toward our education goals. Instead of just displaying the Champlin collection of fighter planes, we were telling the stories of the people who built and flew them as well as the historical events in which they played a vital role. We also discovered that the museum could educate adults as well as young people. I was amazed at the general lack of understanding about how the events of the first half of the last century impacted the world today.

The "blockbuster" exhibit *Leonardo da Vinci: Man, Inventor, Genius* was a huge success and demonstrated not only that we could manage a major event, but more importantly, that similar exhibits could attract a broader and different audience.

Over the last 10 years, we have continued to expand our commitment to STEM education. The Washington Aerospace Scholars program, the Aviation Learning Center, and supporting the opening of the Raisbeck Aviation High School on our campus reflect the trustees' commitment to education. More recently the establishment of the Alaska Airlines Aerospace Education Center and the Boeing Academy for STEM Learning has confirmed the museum's leadership in education. As the father of two daughters, with two granddaughters, I am particularly proud of our outreach to young women with *Women Fly* and the programs built around Amelia Earhart.

In 2010, I was asked by the board to serve as interim CEO as we began a search for a new CEO. It was a particularly challenging time as I had to deal with impacts of the recession, including having to restructure our banking relations and manage some budget issues, as well as address low staff morale that had resulted from pay freezes and increased workload. On top of that, the board had committed to compete for one of the three retired space shuttles. To be responsive to NASA's RFP, we had to have a building that could house a shuttle and conduct an extensive marketing effort to be successful. I did get a few quizzical looks from the usually conservative board when I told them we were going to have ads on buses and fly our shuttle flag from the top of the Space Needle. Despite the challenges, with terrific support from the board and staff, we finished the Space Gallery in record time.

I look at 2010 as a major milestone in the history of the Museum of Flight. We got our financial house in order, hired Doug King, a highly experienced science center CEO, significantly improved the morale of our terrific staff, and built the cornerstone of our west side campus. While we did not win a flown shuttle, the Full Fuselage Trainer that is now in the gallery has turned out to be a far more effective teaching tool. We also started a strategic planning process, Vision 2020, that set the course for the next few years and is supported by the $77-million-dollar *Inspiration Begins Here!* campaign.

All the amazing things that have been accomplished at the museum could only happen with the hard-working and dedicated trustees, staff, docents, and other volunteers as well as people from the area who have come to believe that the Museum of Flight plays a critical role in the life of this community. If I tried to name them all, I would not know where to stop and would run the risk of leaving someone out. However, we all should be very thankful for the commitment that Bill Boeing Jr. made to the museum over many years. From moving the Red Barn to our campus to his campaign gifts, his financial support was tremendous. Bill's contributions were much more than financial. I valued the common-sense advice and perspective on the impact of aviation on our world that he communicated through his stories, typically over a sandwich at his kitchen table. He was very proud of what the museum has accomplished and will accomplish.

— Mike Hallman

The museum's
Boeing B-29 is moved
from Boeing Plant 2
to a hangar during
the restoration
process, 2010.

CHAPTER EIGHTEEN

Special Performance

New CEO Douglas R. King assumed his position in January 2011, after 15 years as CEO of the St. Louis Science Center, one of the largest of its kind in the country. He knew the city from his graduate work in finance at the University of Washington, and he also knew the museum from his former position as president of the Challenger Center for Space Science Education and participation in technical and museum associations.

King was attracted to the forward thinking of the museum's board and the quality of its educational programming. As he had run a large science and tech attraction that annually served more than a million visitors, the scale of operations was not daunting. He was eager to finish some of the projects at hand and begin a new planning process for the next 10 years. King wanted to put a shine on all components of the physical plant, programs, and events.

With an established national museum leadership role and a valuable community asset, the foundation's challenge was to continue growth and expand excellence. It was time for an assessment of both long-range and strategic planning. King and the board revisited Vision 2020, a commitment to a package of goals and objectives initiated by the board in 1997. The upgrading process involved all functional committees with an agenda that encompassed education, collection, exhibits, restoration, visitor services, and the facility and operations. This planning also incorporated concept designs for improvements and cost estimates for capital requirements and program support.

Kevin Callaghan was finishing his second year as chair of the board, and he and others looked to the new CEO and his considerable museum experience as a stabilizing influence. In this instance, stabilizing does not mean standing still, which is not possible in such a fluid environment of collecting, interpreting, educating, and entertaining. Growth and change are a constant force.

The museum had matured from pioneer days into a bustling, sophisticated, world-class campus. It was bigger, better and, correspondingly, more expensive to operate. The planning for the future had to incorporate capital fundraising along with annual support to operations. King was experienced in the major museum dynamic and suited to the role. He had his work cut out for him.

King established a multiyear planning process to assess all functions of the museum operation, with a focus on the collection, public programs, and the visitor experience. This comprehensive evaluation resulted in a revised roadmap for the next decade that addressed the priorities in development and allocation of resources. This long-range vision feeds into annual operating budgets. Working with board and committees, King integrated the planning with fundraising, again tailored for phased development. Funding became a package to include the capital needed for new structures, the budget for remodeling and improvements, money for educational programs, and annual support to operations.

Emphasis on space subjects continued, with NASA Administrator Maj. General Charles Bolden's visit in 2011 as part of the Michael P. Anderson Memorial Aerospace Program series. Taking advantage of NASA's downlink capability, the museum joined in live communications with the crew at the International Space Station.

The esteemed B-29 bomber restoration volunteers received well-deserved attention from the Association of King County Historical Organizations (AKCHO) in April 2011. In a ceremony held at Seattle's Museum of History and Industry, Project Manager Dale Thompson was honored with the Willard Jue Memorial Award, and the team earned the Long-Term Project Award. More

than 70 people had donated thousands of hours of their time to ensure that the plane was brilliantly restored. Boeing company historian Mike Lombardi had nominated the group, stating: "The B-29 Team has done what I consider a miracle; they have transformed what was essentially a trash heap of oxidized aluminum and rusted steel into a shining B-29 Superfortress appearing as it once did when it rolled off the Boeing production line more than 65 years ago."

The traveling division of Seattle's 5th Avenue Theater brought its production of *Rosie the Riveter* to the museum in April 2011. The Adventure Musical Theater

Top: Live communications with the crew of the International Space Station, 2011.

Bottom: B-29 restoration volunteers, April 2011.

Touring Company performed the play, written and composed by local Northwest talent, which showcased the women who joined the workforce to help build planes during World War II. Continuing on that theme, in July 2011 the museum presented "Heroes of the Homefront: A Celebration of the B-17," with displays, tours, and an after-hours hangar dance in the museum's side gallery.

August 2011 introduced the Red Barn Heritage Award, an occasional recognition of great contributions to the museum and aerospace community. William E. Boeing Jr. was the first recipient, and he was a popular choice. A large gathering of friends and colleagues joined the honored guest and his wife, June, on the occasion. The attendees frequently applauded the expressions of appreciation accompanying the award. While enumerating all of Bill Boeing's contributions over the years to the museum, financial and otherwise, speakers pointed out that perhaps the most important was pitching in to save the historic Red Barn. This recognition was appropriate in that it formally identified the historic factory building as a symbol of community service. In acceptance, Bill Boeing thanked all who had supported the project over the years, saying again, as he often did, "It's the people, not the planes" that make the institution.

The Aviation High School, after a decade of teaching in interim facilities at South Seattle Community College and a former Highline elementary school, received its desired location near Boeing Field and the Museum of Flight.

On August 23, 2011, the school held a groundbreaking ceremony on the property made available adjacent to the west side museum campus. Upon success of the private fundraising that complemented the public investment in the school, and with a property lease from the museum, the new $44.5 million building was ready for construction. Principal Reba Gilman explained that "from the school's inception, the museum has been envisioned as its eventual home. This location will provide ready access to the museum's resources and put the school on the doorstep of some 200 flight-related businesses operating out of the Boeing Field area."

Top: Mike Hallman, trustee and former Interim CEO of the Museum of Flight (right), presents the Red Barn Heritage Award to Bill Boeing Jr.

Bottom: Aviation High School students, Principal Gilman, and local dignitaries at the groundbreaking of the Raisbeck Aviation High School on the Museum of Flight campus, 2011.

B-29 Bomber

By far the most complex aircraft of World War II, the Boeing B-29 Superfortress entered service in 1944 with performance not dreamed of when that conflict began five years earlier. This U.S. Army Air Forces bomber—the most capable of WWII—earned lasting fame in the Pacific Theater of Operations and soldiered on with the U.S. Air Force in the postwar era.

With vast expanses of open water to cover in the Pacific Theater, obtaining a more capable bomber was a top military priority. The B-29 program would in fact be the single most costly weapons project undertaken by the United States during that conflict, exceeding in cost the Manhattan Project, with which this aircraft type would forever be linked. The B-29's accelerated development and fielding in fact marks one of the most dramatic home-front successes of the war.

It all began in December 1939, when the USAAF issued a formal specification soliciting the creation of a "superbomber" featuring great range, very fast cruise, and a bomb load many times that of the B-17 Flying Fortress. Boeing responded the following May with the Model 345,

Tricycle landing gear and two tires per strut accommodated the B-29's huge operational gross weight.

a design much more advanced than competing proposals submitted by Consolidated, Douglas, and Lockheed.

Judged the winner, Boeing received an Army contract in August 1940 to develop a very heavy bomber to be designated the B-29. Consolidated's less ambitious proposal, which would become the B-32 Dominator, was likewise funded as a precaution in case the ambitious B-29 failed to materialize.

Because B-29s could fly missions up to 18 hours long at altitudes approaching 32,000 feet, the company realized that its new bomber needed to be pressurized. This led the design team to adopt a round fuselage with a blunt nose lacking the usual stepped cockpit. A tunnel through the unpressurized bomb bays would allow in-flight passage between the forward and aft crew compartments.

No existing wing could provide the needed combination of lift, drag, and airfield performance, so Boeing further improved an airfoil it had developed for the XPBB Sea Ranger, a large patrol flying boat prototype built for the U.S. Navy. To this superb wing, the B-29 design team added powerful Fowler flaps to keep takeoff lengths at gross weight within acceptable limits.

To accommodate wartime gross weights at times exceeding 140,000 pounds (63,503 kg), Boeing's design employed a tricycle landing gear with two tires per strut. Power was provided by four Wright R-3350 Duplex-Cyclone radial engines, each rated at 2,200 hp. Still being developed, these troubled engines—which subsequently achieved reliability and greater power—were the single greatest source of program risk and frustration.

Boeing workers had already contributed 1.3 million labor-hours to the B-29 when the first prototype took to the air on September 21, 1942. The following February, the second prototype suffered a fatal crash as a result of a catastrophic engine fire, necessitating rapid revisions to the R-3350 engine.

So crucial was the B-29 program that production commenced at four locations: Boeing Renton in Washington, Boeing Wichita in Kansas, and licensed production by Martin near Omaha, Nebraska, and Bell in Marietta, Georgia. Aircraft rolling off these assembly lines flew directly to modification depots, where an ongoing flood of design revisions had to be incorporated before they could be delivered. Since many depots lacked hangars large enough to accommodate Boeing's Superfortress, work was often performed outdoors in freezing winter conditions. Ultimately successful, this Herculean delivery effort is today remembered as the Battle of Kansas.

Boeing B-29s entered combat in June 1944. Initial operations were from China and India, until bases were secured in the Mariana Islands. Following the capture of Tinian, Saipan, and Guam, these islands became home to vast fleets of Superfortresses, with hundreds of airplanes often dispatched at once. With the capture of Iwo Jima, halfway to the Japanese mainland, U.S. forces secured an interim base for B-29 emergency landings and use by escort fighters and air/sea rescue planes.

On August 6, 1945, the Boeing B-29 *Enola Gay* dropped an atomic bomb on Hiroshima, Japan. Three days later, the B-29 *Bock's Car* dropped another nuclear weapon on Nagasaki. These cataclysmic events brought World War II to a close. In the postwar era, B-29s were joined by the B-50, a much improved Superfortress distinguished by its 3,500-hp Pratt & Whitney R-4360 engines and taller tail. These close cousins were the mainstays of U.S. strategic airpower until superseded by swept-wing Boeing B-47 Stratojet and B-52 Stratofortress bombers in the jet age.

Leveraging its world-leading technology, Boeing introduced the wartime C-97 Stratofreighter, a pressurized cargo plane combining a fatter fuselage with the wings, tail, and systems of the B-29 or B-50. In the postwar era, USAF C-97s and KC-97s served in large numbers as transports and aerial refueling tankers, respectively. In 1949, Boeing also introduced the 377 Stratocruiser, which is fondly remembered as the most capable and comfortable of the great four-engine piston airliners.

On October 14, 1947, a modified B-29 flying high over the California desert drop launched the Bell X-1 rocket plane *Glamorous Glennis*, allowing USAF Captain Charles E. "Chuck" Yeager to shatter the sound barrier. It was a fitting mission for a great bomber that itself did so much to advance aviation.

The flight deck of the B-29.

The Number 3 Boeing 787-8 on the tarmac.

Over these last few years, progress at the museum continued apace, with steady growth in public service, earned income, and educational programming, and the expansion of the physical plant. While introducing a reevaluation and new course, the museum maintained its exemplary mission in flight museum collecting, interpretation, research, public programming, special events, and both formal and informal education. It is abundantly clear that the MOF is making a difference in math and science literacy with its variety of STEM programs both in-house and through outreach.

Exhibits staff just trucked on with their new installations and constant upgrades. The record-setting Perlan Glider flown by the late Steve Fossett and the Sikorsky HH-52 Seaguard helicopter moved into the Great Gallery in 2011. A "Number 3" Boeing 787 Dreamliner, a recent member of the Boeing fleet, went to temporary outdoor display in 2014.

King and the museum board organized the *Inspiration Begins Here!* Campaign. The big item in this package was the Aviation Pavilion. SRG Architects designed the initial concept to cover the large aircraft, which was estimated to cost more than the foundation board thought they could raise. The funding feasibility study identified a number that would require some trimming of the facility. Ed Renouard again was confronted with a reduction in the budget and scale of the structure.

The first phase, to house five or six aircraft, was estimated to cost $25 million. As an alternative, and to accommodate more aircraft, Renouard developed a concept he called the "aircraft carport," a larger building with a roof and no siding—a simple enclosure to protect the planes from the weather. The structure was modest in cost for its size, while still providing a safe environment for the artifacts and much more comfortable space to welcome and educate visitors. The development team took on a big challenge and, with imagination, turned it into a new inspiration for the institution.

The display berm outside the entrance was loaded with planes, so the museum began using the site to the west of the main campus for the larger aircraft such as the 747, Concorde, and Air Force One. During planning for the Aviation Pavilion, in Summer 2012 the Airpark was upgraded with visitor amenities.

In late May 2012, an opening ceremony introduced the Bell UH-1H "Huey" helicopter to the Great Gallery. This artifact, donated by King County, had required a lengthy restoration, and that team joined a group of museum and public officials, Vietnam veterans, and the public in a program of respect. Speakers featured Maj. General Patrick Brady, U.S. Army Retired. General Brady, a recipient of the Medal of Honor, was recognized for his courage flying Huey air ambulances in Vietnam. His experience with the incredible DUSTOFF helicopter rescues of wounded soldiers is another chapter in the annals of American ingenuity and bravery. General Brady pointed out that the Huey flew more combat than any other aircraft in history. Again, the museum proved its technique of interpreting historical aircraft with personal stories of service and courage.

The "Wings of Heroes" gala on September 22, 2012, had one of the largest turnouts ever. Co-chaired by Mike Hallman, Bruce McCaw, and Charles Simonyi and organized by Bill Rex, this extraordinary event was the result of a year of planning. The evening featured 40 astronauts and space pioneers, from early missions to current, at hundreds of tables in the tented annex to the Space Pavilion and Airpark. Among those attending were Buzz Aldrin, Dick Gordon, and Jim Lovell, and returning again to the museum, cosmonauts Alexei Leonov and Valeri Kubasov. The list of space luminaries also included legendary flight director Gene Kranz and other NASA pioneers. Private air and space companies were also prominently represented.

Veterans and the aircraft restoration crew gather at the installation of the Huey in the Great Gallery.

Huey

No Western helicopter series has played more roles, worked harder, or been built in more varieties and greater number than the legendary Bell UH-1 and its descendants. Introduced to service by the U.S. Army in 1959 as the HU-1 Iroquois, it became iconic in the 1960s under the revised designation UH-1 and the nickname Huey.

In the latter 1950s, the Sikorsky H-34 was the Army's standard aerial workhorse. A large piston-powered helicopter, it could transport heavy loads but was cumbersome, slow, and ungainly.

The museum's Bell UH-1H Huey before restoration, circa 2003.

The Army in particular needed a nimble utility helicopter, but procurement funds were available only for a medical-evacuation type, so it was in this role that the Bell Model 204 came into being. The first prototype took to the air at Bell's Fort Worth plant on October 20, 1956.

The Huey was the U.S. military's first turbine-powered rotorcraft. Exploiting turboshaft power, Bell engineers placed the Huey's compact engine and transmission above a broad cabin with wide doors at both sides. The result was a compact, low-slung, unencumbered deck that could easily be loaded and unloaded anywhere in very little time. Bell's design also provided superb visibility, excellent handling characteristics, and a robust two-bladed rotor system with high inertia for safety.

The most successful helicopter in U.S. history, the Huey would soon surpass the total number of all the world's previous helicopters combined. Sleek and trim, it represented a dramatic leap in flexibility and performance, enabling transformative rotary-wing operations. In widespread service by the early 1960s, the UH-1 met every challenge placed before it as the decade progressed and U.S. military involvement in Southeast Asia escalated.

Hueys flew with all four U.S. services during the Vietnam War. In Army and Marine use, they distinguished themselves in the light utility, observation, gunship, rescue, and casualty-evacuation (CASEVAC) roles. Although combat and attrition together would claim a large number of Bell UH-1s, the type proved as indispensable in Vietnam as the Jeep did in World War II.

CASEVAC operations returned the Huey to its roots. Known less formally as DUSTOFF missions for the radio call sign under which they were performed, they put this helicopter's speed, ruggedness, and resilience under fire to the ultimate test. DUSTOFF crews routinely risked all to rescue the wounded and race them to field hospitals when every minute was precious. To this day, countless thousands of veterans feel profound gratitude to the Huey and its daring crews for saving their lives.

Back at Fort Worth, meanwhile, Bell engineers were adapting their world-leading technology to meet additional requirements. When the Army asked to transport more troops and supplies per sortie, Bell lengthened the Huey and gave it uprated power to create the UH-1D (Model 205), a "squad carrier" that could seat 15 troops or carry six stretchers at once. Ongoing improvements and further power increases saw the D-Model evolve into the definitive UH-1H by 1967. Bell delivered more than 4,850 H-models to the U.S. Army alone, making this the most produced Huey variant.

As if this weren't enough, Bell used its own funds to secretly develop a dedicated attack helicopter. With so many orders already on its books, the Fort Worth company was not allowed to bid for this new business, but Bell's leaders guessed other manufacturers might experience developmental delays. When this proved to be the case, Bell unveiled the AH-1 HueyCobra, which combined the dynamic system and other Huey components with stub wings for ordnance and an all-new fuselage with stepped cockpits for its pilot and gunner.

But this wartime proliferation of models was just the beginning. In the decades since the Vietnam Conflict, the UH-1 has sired an amazingly diverse stable of civil and military offspring. The twin-turboshaft Bell 212, Bell 214 Super Transporter with 18 seats, and Bell 412 with its four-bladed rotor are among the many Huey variants serving operators in demanding roles and rugged environments around the world. Civil models have long been built in Canada. Many other countries have also produced civil and military Hueys under license over the years, bringing the total number built to more than 16,000.

In service around the globe, Hueys and HueyCobras have amassed more combat flight-hours than any other aircraft type, be it fixed or rotary wing. Although long since superseded in first-line Army service by the Sikorsky UH-60 Black Hawk, the Huey is again in production for the U.S. Marine Corps. Today's state-of-the-art version is the UH-1Y Venom. The Marines are also taking delivery of new Bell AH-1Z Vipers, a fully updated HueyCobra for the 21st century.

With such a remarkable string of accomplishments, it's little wonder that Bell's Huey family is widely regarded as the most significant helicopter series in history.

The Huey restored and on display in the Great Gallery.

Emcee Steve Pool, museum board member, television weatherman, and pilot, called up the celebrities in a long introduction for the grand evening. The guests ranged from first space traveler Joe Engle, the pilot of the X-15, all the way to the modern space tourist Simonyi and entrepreneur Jeff Bezos. Also attending were father and son Owen and Richard Garriott, the former a veteran of Skylab and the latter a member of the crew of the International Space Station.

A highlight of the evening was Martha Chaffee's presentation of Deke Slayton's cherished gold-and-diamond astronaut pin to the museum. Martha Chafee is the widow of Roger Chaffee, who in 1967 perished in a preflight launch test of Apollo 1 along with Virgil "Gus" Grissom and Edward H. White.

The story of this pin explains its value to the collection. Pins are standard in the flight profession. Wally Schirra promoted such an award for the Mercury 7. NASA agreed and offered a silver pin upon selection to the corps and a gold pin after a first spaceflight.

Deke Slayton picked the crew for the first Apollo mission. Deke, temporarily grounded by a heart issue and serving as Director of Flight Crew Operations, selected Gus Grissom, Ed White, and Roger Chaffee. These fellow astronauts knew that their friend Deke was unlikely to get a chance to fly, and upon urging from Grissom, presented him with a ceremonial gold pin. Knowing he would be reluctant to accept, they had it designed as a one-of-a-kind with a diamond inset. They never had a chance to present it formally, as all three were lost in the fire. In a touching tribute, the widows of the three lost astronauts gifted the pin to Slayton after the tragedy. Deke was so touched by this gesture that he wore the pin daily. It became to him, and to others, a wonderful testimonial to the collegial spirit of the astronaut corps.

Deke Slayton's gold-and-diamond astronaut pin.

Later, Apollo 11 commander Neil Armstrong asked Slayton for the pin "for a few weeks" to take into space with other mementos. Thus, this pin traveled to the Moon and back in 1969, lending even more meaning. In 1972, Slayton was returned to flight status and assigned to the Apollo-Soyuz mission, which succeeded in 1975. Slayton earned his gold pin, but it was the diamond pin that he continued to wear.

Deke Slayton was a big fan of the MOF, participating in several events, including the 15th-anniversary commemoration of Apollo-Soyuz. As well, Neil Armstrong respected the work of the museum and suggested that the museum was a good place for this incredible artifact. The widows of the original astronauts agreed.

Armstrong died the month before the "Wings of Heroes" event. His sons, Rick and Mark, paid tribute to their father, which touched all in attendance. Neil was a national hero and also a longtime supporter of the MOF. His presence, so missed, was felt that evening as Martha Chaffee presented the pin to the museum on behalf of the Apollo 1 astronauts. At times it can be the small items that make the biggest impact, and the Slayton pin helps to tell the space story as well as the much larger artifacts. The evening's production manager, Doug Tolmie, said it best: "The gala validates the Museum of Flight's clear position as one of the finest storytelling institutions in the world."

The 2012 "Wings of Heroes" gala celebrated the first 50 years of space flight, and honorees included astronauts, cosmonauts, flight controllers, and NASA personnel.

The star-studded crowd experienced an evening that none will forget. The accolades were immediate and continued. One attendee, Amnon I. Govrin, a software engineer and space enthusiast new to the community, wrote extensively of the gala on his blog site. He explained how incredible it was to be among so many individuals from the 50 years of space history, from the Mercury 7 and Apollo astronauts through to the Space Shuttle. He expressed his delight at mingling with representatives of all these missions up to the current private initiatives. Govrin concluded, "This was a once in a lifetime event. A group of so many who trail-blazed human spaceflight, a list so big and spanning all of the history . . . We also did our small part, contributing to a wonderful museum that plays a part in inspiring and educating what could become the next generation of astronauts, scientists and engineers."

The evening created lasting memories, raised serious money—some $2.1 million—with its auction, and elevated the image of the institution. Museum chair Hallman summed up the importance of the evening: "Manned space travel is a book without end. The museum is telling the story of manned space flight, past, current, and future."

In November 2012, the Full Fuselage Trainer was opened to the public in its new Space Gallery home, with first tours for the Washington State Governor and other dignitaries. With the Space Gallery, the museum notched up its education offerings. The price for transportation and reassembly of the FFT was $1.85 million, with another $1 million invested in access improvements and exhibitry, but was well worth the cost as it quickly became a museum centerpiece that lived up to the promotion that "Now everyone can be an astronaut."

The FFT has become a powerful attraction. Its long life as a trainer for astronauts translates dramatically for an educational experience. The FFT is big, 122 feet long and 46 feet tall. Transporting it and putting all of the 22 pieces together was a big chore, and that was just the beginning. The exhibits team added lighting, a sprinkler system for fire suppression, and other improvements for access that include an elevator for the physically challenged.

Neil Armstrong

Neil Armstrong (second from left) with Bill Boeing Jr., Bonnie Dunbar, and Bruce McCaw.

On July 20, 1969, Neil Armstrong became the first person ever to set foot on another world. "That's one small step for man, one giant leap for mankind," he said as he stepped onto the lunar surface, an event breathlessly watched around the world.

The flight of Apollo 11 and those that followed rank among the most daring voyages of scientific discovery in history. Twelve astronauts left footprints on the Moon before the Apollo Program ended in 1972. While untold thousands of people worldwide contributed to its success, Armstrong's name will forever be most closely associated with human beings taking their first step off their home planet.

Ironically, few men were ever less desirous of fame than Neil Alden Armstrong, a shy, soft-spoken Ohio native. Born in the town of Wapakoneta, he fell in love with flying when his father took him to the National Air Races at age two and treated him to his first airplane ride three years later in a barnstorming Ford Tri-Motor.

Young Neil built and flew model airplanes. Working at the drugstore, he earned enough money for flying lessons at a grass airstrip, soloing at age 16. He also loved scouting and became an Eagle Scout.

In 1947, the teenager entered Purdue University to pursue aeronautical engineering. Interrupting these studies, he trained as a U.S. Naval aviator, winning his wings of gold in 1950, just as the Korean War began. Flying from the deck of the USS *Essex*, he completed 78 missions over Korea in Grumman F9F-2 Panther jet fighters. One mission saw his Panther badly damaged by flak, forcing him to eject.

After completing his degree, he went to work for the National Advisory Committee for Aeronautics (NACA), which became the National Aeronautics and Space Administration (NASA) in 1958. Serving as a civilian engineering test pilot, he flew a wide variety of research and other aircraft. This activity culminated in seven flights in the hypersonic North American X-15 rocket plane.

Armstrong applied to NASA's Astronaut Corps in 1962 and was accepted. Aboard Gemini 8 four years later, he and crewmate Dave Scott successfully completed the first-ever docking of two spacecraft in orbit. Disaster struck when the spacecraft suffered a stuck thruster, which caused it to tumble wildly in orbit. But for Armstrong's skillful piloting and coolness under pressure, they might have been lost.

Assigned to Project Apollo, Armstrong commanded his second and last spaceflight aboard Apollo 11 with fellow crew members Edwin "Buzz" Aldrin and Michael Collins. Boulder-strewn lunar terrain prompted him to take over and use up precious fuel before setting down safely in the Sea of Tranquility. "Houston, Tranquility Base here," Armstrong transmitted. "The *Eagle* has landed."

"Roger, Tranquility, we copy you on the ground," came the voice of fellow astronaut Charlie Duke, serving as CAPCOM for Mission Control. "You got a bunch of guys about to turn blue. We're breathing again. Thanks a lot."

Never one to rest on his laurels, Armstrong had pursued a master of science degree in aerospace engineering, which he received from the University of Southern California in 1970. The next year, he left NASA to begin teaching aerospace engineering at the University of Cincinnati, retiring finally in 1979.

Armstrong and his family were very private. For years, he personally answered letters from young fans, spending hours a day at this activity. Guided by his own characteristic humility and personal integrity, he refused to make any endorsements or attend exploitive events, but appeared at public events if they supported worthy causes and emphasized others over himself.

A strong supporter of the Museum of Flight, Armstrong visited many times to participate in space-related celebrations, such as the museum's Friendship One, Apollo 13, and X-15 events. So much did he love the museum that he even arranged for his Korean War fighter squadron to stage a reunion here.

Neil Armstrong died in 2012 at age 82. His family, friends, colleagues, and students fondly remember him as a superb engineer, patient teacher, and one of the finest pilots ever to fly. To the rest of us, he seems larger than life, a figure to be admired and wondered at for the skill, intelligence, and courage he brought to one of the most complex and audacious collaborations in human history.

NASA Director Charles Bolden, himself a former military aviator and astronaut, perhaps said it best: "As long as there are history books, Neil Armstrong will be included in them, remembered for taking humankind's first small step on a world beyond our own."

The 60-foot payload bay is used for visitor assembly, and guided tours afford views into other smaller spaces. Videos allow for viewing of the entire interior. Interactive devices and interpretive panels reveal the history and technology of the trainer, built in the 1970s at Johnson Space Center Houston. On a somber note, the families of Dick Scobee of Auburn and Michael Anderson of Spokane donated the training suits of each astronaut. Scobee died in the loss of *Challenger* and Anderson in *Columbia*.

In 2012, the museum's education staff traveled internationally to present programs in Qatar and opened up distance-learning curricula across the United States and Canada that garnered a prestigious Pinnacle Award from the Center for Interactive Learning and Collaboration. The Museum Apprentice Program introduced budding young professionals to real-world museum tasks.

Anne Simpson as Amelia Earhart at the 2013 Electra Fly-In.

The Education Department took programs to Malaysia in 2013, continuing its global influence.

Programs and individual museum educators continued to win regional and national awards, with recognition in 2013 from the National Science Teachers Association of an MOF educator. Marketing gained chops, receiving a highly competitive American Advertising Award in 2012 for the imaginative "Now Everyone Can Be an Astronaut" campaign, created to celebrate the opening of the Full Fuselage Trainer in the Space Gallery.

In September 2013, the 1935 vintage Lockheed Model 10-E Electra flew into Boeing Field to serve as the centerpiece of an exhibit to open the following month in the Great Gallery. One of just two of these models in existence, it is identical to the plane Amelia Earhart flew on her ill-fated flight around the world in 1937.

Museum curator Dan Hagedorn noticed the aircraft's advertisement for sale and reported it through the system. At a meeting, chair Hallman mentioned the availability and the price tag, asking if this was something to consider. Board member Anne Simpson responded, "Speaking for 52% of our population, I think it's a great idea." With the agreement of the board, Simpson enlisted fellow commercial pilot and board member Nancy Auth and reporter Patti Payne, and they all went to work. Simpson, elected chair of the board in 2016, had spent years helping to diversify the museum, and the Electra was an excellent opportunity to underscore that effort. She and Auth structured the campaign to be "Amelia-esque," in their words. As an adjunct, Simpson and Auth organized a women's society in support of the museum, the Amelia Loves Her Lockheed Ladies' Guild, or ALL Ladies' Guild. "We expect that women in all walks of life will join us." The Project Amelia campaign met its target of $1.2 million with large contributions by Wells Fargo and Alaska Airlines and additional dollars from 600 individual donors.

Let's Do the Numbers

The Museum of Flight welcomed more than 500,000 visitors for the first time in 2012. Although this attendance has since been surpassed, this is a good year to compare with the early opening of the Red Barn phase 1, the performance for the ramped-up phase 2 Great Gallery, as well as the fast climb in 2004 upon opening the Personal Courage Wing.

Along with this historical visitation, virtually all categories scored high on the chart in 2012. The museum served more than 150,000 participants in its education programs. Membership returned to almost 20,000 individuals and families. Earned income, admissions, gift sales, program fees, and event revenues were up and climbing. A record number of volunteers supported the 150 full- and part-time staff. Earned income generated 64% of the $16 million budget. Assets soared to $128 million.

As the board leadership points out, "It is not just about the numbers," as many assets cannot be easily quantified. The education programs are the best among flight museums and are reaching out internationally. Exhibits are excellent, the collection one of the best. The library and archives have grown to be one of the largest in the world. Volunteer numbers and skills are an unparalleled resource. The qualitative result of all these components delivers exceptional public service to Washington State and the world.

The Lockheed
Electra soars over
the museum on its
final flight.

The aircraft had been restored in 1994 to match Earhart's model and then flown around the world by aviatrix Linda Finch in 1997 to honor Earhart and her aviation contributions. "This rare and remarkable aircraft will be more than an addition to the Museum of Flight's world-class collection," said Simpson, a Delta Air Lines captain. "The real story here is motivating and inspiring young people, especially girls, to take some risks and become the best they can be."

A display entitled "In Search of Amelia Earhart" introduced the Electra into the Great Gallery in October 2013. The exhibition relates a comprehensive story of Earhart and her many talents and accomplishments as an educator, writer, artist, fashion designer, and feminist. The display also incorporates the only remaining piece of Earhart's personal Electra, salvaged from her first attempt at an around-the-world flight in March 1937, when she ground looped the plane on takeoff from Ford Island in Hawaii.

The Electra has connections to other flight stories in the museum. In its commercial model, it was a competitor with the Boeing 247 and Douglas DC-2, both in the museum collection. The Electra also relates a story of legendary designer Clarence "Kelly" Johnson, who redesigned the aircraft as one of his first projects with Lockheed. Johnson's work is well represented in the museum, including the Blackbird nearby in the Great Gallery.

Of interest to all, the exhibit is particularly motivating to young women. As an adjunct, the "WomenFly!" program is an annual event pairing mentors with students in middle and high school. The education programs continue to identify areas of need and to design programs to serve those students.

On October 17, 2013, Raisbeck Aviation High School (RAHS) celebrated its grand opening. The school had already established a stellar reputation, rated in one national review as the sixth

best school in the state. Now, it was preparing to move its 400 students into the new facility designed to accommodate the STEM curriculum with a focus on aerospace. Education, industry, and museum representatives joined with students and faculty to celebrate the beautiful building, located on the west campus of the Museum of Flight.

RAHS is a proven success, with 95% of graduates going on to college. The school, operated by the Highline School District, draws half of its student body from that area, with others from around Puget

Lieutenant Commander Nora Jacobsen VR-61 instructs students at a 2012 "Women Fly!" Event.

Sound. The competition for entry is keen, as so many young people are attracted to the aerospace STEM curriculum. The new structure and aerospace environment are expected to add to this student enthusiasm and academic success.

The museum experienced yearly upturns in performance, welcoming 540,000 visitors in 2013. Some regional attractions post larger numbers, but not in an out-of-the-way location. This level of visitation and the high quality of the experience are exceptional for an educational institution in an industrial setting. It is the kind of performance hoped for in early planning, now accomplished and with lots of reasons to be confident about continued growth and service.

Education programs continued to attract gifts and sponsors. In April 2014, Alaska Airlines announced a $2.5 million grant to create an Aerospace Education Center. This resource was designed as a one-stop place in the Great Gallery where students, parents, and teachers could easily access educational materials. The announcement mentioned that the Alaska Airlines Aerospace Education Center would be "designed to create pathways to science, technology, engineering, and math (STEM) careers among young people."

Raisbeck Aviation High School

Early in museum operations, Lovering began discussions with South Seattle Community College to develop a formal relationship with aerospace technology and training. This collaboration resulted in joint programs but fell short of an official educational arrangement. This was just the beginning of a search for partners. What was to come later was a national model institution for air and space education.

Principal Reba Gilman speaks at the Raisbeck Aviation High School groundbreaking.

Raisbeck Aviation High School is a full four-year educational facility located on museum property to the north and west of the main campus. It operates as a partner and represents a fascinating step in educational outreach.

The idea for this high school began with Reba Gilman, a teacher, administrator, and experienced educator. During her professional tenure, Gilman taught in several parts of the state, with a specialty in marketing, business, finance, and entrepreneurship. Earning her educational administration credentials at Central Washington University, she served as a vice principal and vocational director. Her position as principal and director of the Sea-Tac Occupational Skills Center in the Highline School District served as a launchpad for her approach to merging school curriculum with job skills. Gilman became convinced that a technology-based preparatory high school with an aerospace focus could engender the best in expectations for Science, Technology, Engineering, and Math (STEM) initiatives.

In 2002, Gilman began her work on developing and marketing the concept of an Aviation High School. She found a dedicated champion in Dr. John Welch, superintendent of the large and diverse Highline School District. Dr. Welch's support provided the base needed. Setting out small, Gilman found a partner at Boeing Field, interest at the Museum of Flight, and later support from South Seattle Community College. With an early gift from the Gates Foundation, she forged all of these partnerships into a series of locations and facilities that allowed her to test her model and prove its worth. Opening in 2004, the school was first located at South Seattle Community College before settling into facilities at Highline School District's former Olympic Elementary School.

Gilman credits Dr. Bonnie Dunbar as the inspiration behind the acceptance of the new school on the museum campus. An advocate of STEM and a hero to students, Dr. Dunbar had a personal story that underscored the potential of this kind of curriculum.

The initial proposal to partner with the museum was presented to the board, then under the leadership of Chair Michael Hallman. The discussion of how to partner with a formal high school and its school district management was new to the museum community, but the foundation agreed to support the innovative school.

Board member Bruce McCaw quietly entered the picture with a strategy to ensure that the Aviation High School could be built to the size and scale required. McCaw was aware of an available private property adjacent to the west side museum campus that the museum wanted and needed for additional expansion and parking. An advocate of the Aviation High School, McCaw saw the opportunity for a partnership and, with this particular property, an adjacency that could efficiently share parking. He concluded that this was a one-time opportunity that just could not be missed. Although the museum board was focused on other development and funding priorities, McCaw advanced the idea, enlisted support, and launched a move to acquire the additional west side property. The complicated deal was successful.

With a site on a portion of this property, the school organized to solicit capital to complement construction money available from the Highline School District. Fortunately, Sea-Tac Airport management included passionate supporters of the Aviation High School, and the Port of Seattle made the first capital pledge of $5 million to the private portion of the campaign. Initially a

Raisbeck Aviation High School students march to the site of their future school to participate in the groundbreaking ceremonies.

controversial decision, this support was justified by its relevance to aviation and travel divisions, the benefit of the school in training for required job skills, and its related economic impact. With growing public understanding and support, the Port made an additional donation of $5 million. Over time, the Port of Seattle was recognized for this investment and has gained national acclaim for the school. Alaska Airlines took an early interest and made a pledge of $1 million to kick off the private capital campaign. The Alaska contribution was increased to $1.5 million before the school broke ground. Other significant contributions came from the Washington State Legislature, Highline Public Schools, The Boeing Company, and the McCaw Family Foundation.

One of Gilman's students admired the accomplishments of Raisbeck Engineering, located in Seattle, and wrote to James Raisbeck with an invitation to visit the school. Raisbeck was touched, and accepted. During his tour, the student proudly introduced Gilman and other faculty. James was impressed, and he and his wife Sherry made a substantial pledge. He is extremely proud of the school and makes frequent visits to the classrooms to teach and share aerospace experiences. In addition, he has arranged college scholarships for graduates.

The Boeing Field community was an early advocate for this new initiative. Among Gilman's early supporters was Rita Creighton, the community relations manager at Boeing Field, who helped coordinate events and classes on the airfield. Peter Anderson of Galvin Flying in Seattle was another early supporter and became the first chair of the fledgling school's board.

On October 17, 2013, Raisbeck Aviation High School celebrated the grand opening of its new $44.5 million campus to the north and west of the museum. Attendees included Washington Governor Jay Inslee, Boeing Commercial Airplanes CEO Ray Conner, Alaska Airlines CEO Brad Tilden, and James Raisbeck, namesake of the school.

Designed by Bassetti Architects in Seattle, the three-story, 72,000-square-foot structure accommodates the project-based instruction and specialized STEM teaching model. The facility includes classrooms, labs, and maker spaces. The flexible commons area incorporates a cafeteria, sports court, and collapsible tiered seating. The energy-efficient structure features a layout and look that fits into the museum campus and airfield environment.

RAHS is one of the first aviation high schools in the country and the first associated with a flight museum, cementing the role of the MOF as the acknowledged leader in education. The partnership provides the students with dramatic resources. The museum's collection of artifacts, archival materials, and broad base of programs, combined with the airfield location and friendship with the aerospace industry, add to the richness of this school experience. In turn, the school and its committed students increase the value of the museum.

Teachers and students in action at the newly finished school on the Museum of Flight campus.

Raisbeck Aviation High School continues to attract motivated students and provide exceptional learning opportunities. In its third year near the museum campus, *U.S. News & World Report* rated RAHS as the best school in Washington State, and Number 132 in the country. The student body of 420 at this time is 66% male, 41% minority, with 18% economically disadvantaged, and its graduation rate is 98%. RAHS is a good partner and a model for high-school education.

The *City of Everett* is reintroduced in 2014 after an extensive restoration.

The Aerospace Education Center opened in 2015 and introduced 3-D printers, crystal radios, and robots. Improvements to the Aviation Learning Center and Challenger Learning Center were included in this expansion. Additional staff were trained to counsel and advise those using the sources of information, serving as "the nexus for all education-related activities at the MOF."

The precious RA001, the very first 747, was overhauled and reintroduced to the museum in 2014 after 16 years of storage and occasional restoration. The exterior of the *City of Everett* was washed, sanded, and painted in her original white, red, and silver livery. Technicians also re-created the logos of 30 airlines, many no longer extant, that were displayed on the "Queen of the Skies" when she first rolled out of her gigantic hangar in Everett, Washington, on September 30, 1968. It was a grand reintroduction, but it did not appropriately tell the long story of how the aircraft got to looking so good.

The restoration of aircraft is a tedious and expensive process, and the MOF is gifted with abundant talent for this work from aerospace volunteers and others retired in the region. No regional flight museum can pretend to restore and conserve these large artifacts without a loyal volunteer cadre. The 747 restoration is indicative of this asset. To say the sprucing up of the 747 was a big job is not adequate. It was more than that. In the assessment of curator Dan Hagedorn, "No museum, anywhere, has ever faced a restoration project of this magnitude."

Once the construction of the Aviation Pavilion was made a priority, there was renewed energy to get the large aircraft ready for the show. At an MOF meeting, Dennis Dhein and Ted Schumaker volunteered to take on the 747 project. Retired from Boeing, these two had previously worked on the crew of the B-29 restoration. They immediately recruited other volunteers. In late 2012 they pried open the cabin door to find that they had bitten off a lot. The aircraft, sitting idly, had suffered interior deterioration from water seepage. What is a bare interior for the test bed was a mess. Though shocked, the crew was undeterred.

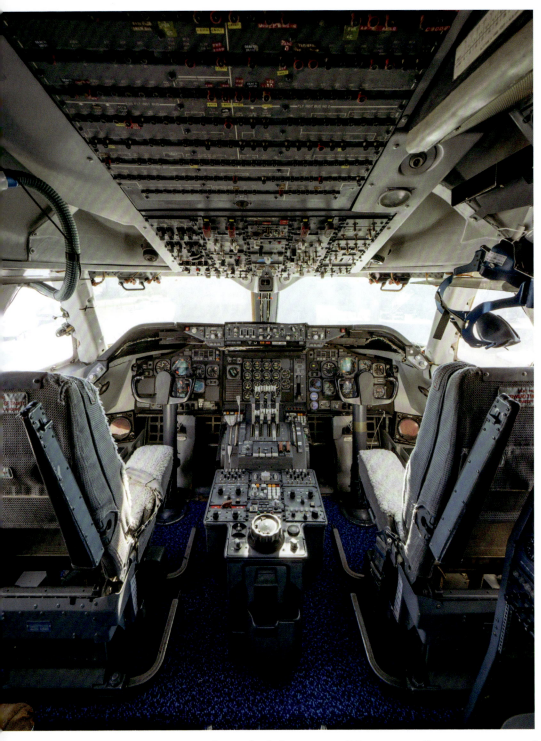

Cockpit of the RA001.

Crew chief Dhein and volunteers began an assessment and task list that ultimately enumerated 118 items. Parts are not easy to find for the first of a kind. One huge complication is that the team cannot expect to find engineering drawings for a plane that still has models in production. Many other obstacles were magnified by the sheer size of the aircraft. It was more than a big job. Chief Engineer Joe Sutter and his original 747 team were named "The Incredibles" for their vast project. The new team of restorers earned this title for their commitment.

It is impressive to see what this determined crew did in just a few years to make the great aircraft presentable for the Aviation Pavilion. It has been cleaned up, with significant parts restored, and given an authentic paint job. Although they still have work to do—at least 40% of their checklist—RA001 looks shiny and dignified in its new home. Guided tours reveal the massive volume of the interior, now dried and cleaned and well along in being brought back to the aircraft test bed that it was. The crew will continue to work on this restoration until satisfied that it is authentic and true to its Boeing roots. The contributed hours of work and donated parts and pieces and skills will be worth millions of dollars. Again, in the words of Hagedorn, "We have brought that beautiful airplane back from the brink."

The 747 is an attraction that helps to tell a lot of stories of Boeing accomplishment, jumbo jets, and commercial transportation history and technology. It can also tell quite a story about volunteerism and the restoration and conservation of artifacts. Over the next 50 years, this icon will help to unfold stories of what we consider the modern era of travel and transportation. It will remain relevant to stories of aviation, and to the development of the Museum of Flight. Once in museum ownership, the RA001 was leased out for engine testing, making millions for the foundation. In turn, that contract served as assigned collateral for a loan that allowed the museum to purchase its restoration hangar at Paine Field. The 747 is a cherished treasure.

Some of the older artifacts were still finding their way into the galleries. The FM-2 Wildcat arrived from the museum's Restoration Center at Paine Field for display in the Personal Courage Wing. It has taken decades in some cases, but the history was preserved, and the items restored, much as was desired in the beginning.

In November 2014, the Michael and Mary Kay Hallman Spaceflight Academy exhibit was added to the Space Gallery. Surrounding the FFT, the 15,000-square-foot display illuminates the history of spaceflight and its future endeavors. The community instantly embraced the new gallery, which continues to grow and advance the theme of aerospace education. The Hallmans' gift expanded on the educational interpretation of the trainer. Other enhancements are planned as additional space artifacts are acquired.

The Museum of Flight experienced a period of expansion to finish off the first 50 years. It has come a long way, but the team has more to do. Even when excellent, there is the determination to be better, and the passion to be the best.

The museum's FM-2 Wildcat, a former "park queen," took decades to restore.

Boeing 747 RA001

Seattle natives know there's a celebrity in their midst. So do aviation buffs and airline enthusiasts around the world. It's the Museum of Flight's big, beautiful red-and-white Boeing 747 *City of Everett*, which proudly sports the serial number RA001.

Few airplanes in history have changed the world as much as the very first Boeing 747. Before RA001 flew on February 9, 1969, the largest commercial jetliners in the world—Boeing 707s and Douglas DC-8s—had single-aisle passenger cabins accommodating about 140 passengers. When RA001 touched down, a new jet existed that was so different, it would fundamentally transform air travel.

The 747 reinvented the passenger experience with a bright, spacious cabin nearly 20 feet wide. It had two aisles, making it history's first *widebody* jetliner. To those boarding, the 747 seemed more like a place than a plane. Adding to this effect, the Boeing "Incredibles," led by legendary designer Joe Sutter, had moved the galleys and lavatories away from the walls to create center islands. More than just provide more window seats, this innovation defined seating areas with more roomlike proportions, mitigating the sense of being in a long tube.

With per-seat operating costs an astounding 30 percent lower than those of other jets, the 747 also revolutionized airline-industry economics. It is the airplane that democratized air travel, opening it up to the masses. No longer would flying have to be too expensive for people and families of average means.

Like a glamorous movie star, the 747 created a sensation when it entered scheduled service in 1970. The press and public couldn't get enough of Boeing's new giant, dubbing her "jumbo jet" and "the queen of the skies." So right did she look to the public that her creation seemed preordained.

Behind the scenes, it was a far different story. Boeing had launched the 747 when it was already overextended with three other committed programs: the all-new 737, improvements to the 727, and the U.S. Supersonic Transport (SST), which Congress would later cancel. But economic forecasts showed the world needing bigger jets, so—at the urging of Boeing President Bill Allen—the company's board of directors launched the 747 anyway. Boeing had once again bet the farm, gambling its continued existence on the aerial behemoth.

When the 747 was launched, it was universally expected that SSTs would soon enter service and displace subsonic jets on intercontinental routes. Boeing experts forecast that the world's airlines would require only 200 747s or so before that happened. With this in mind, Joe Sutter and his team also designed their new jet to be a great cargo plane since demand for main-deck freighters would remain even after passenger sales dried up.

Of course, SSTs never met the high expectations that the world held for them. As for the 747, it was for decades the undisputed champion of long-haul passenger service and the backbone of global air freight. In both these roles, it offered unmatched range, capacity, and load-lifting capabilities. Although other airplanes have since eroded or trumped its dominance, it continues to sell and has garnered more than 1,500 total orders as of this writing. The latest versions are the 747-8 Intercontinental and 747-8 Freighter.

With such a colorful background and so much history to her credit, RA001 is in a sense the "drama queen of the skies." Her placid exterior reflects huge technological challenges met and solved, creative financing to stay afloat while begging bankers for more time, and the successful construction of an all-new factory in Everett, Washington, to build jets way too big for existing plants. Somehow, Boeing people made it work and pulled a rhinoceros-sized rabbit out of a hat.

RA001 occasionally flew again when Boeing needed her services. In the early 1990s, for example, she flight-tested new engines for the 777. Today, though, her flying days are behind her, and she occupies a place of honor at the Museum of Flight. So much history in such a pretty airplane . . .

Early flight test of RA001.

Joseph F. Sutter

Joe Sutter was a legendary aeronautical engineer. Hailed as the "father of the 747," he led the jumbo jet's design in the latter 1960s and played key roles in other Boeing programs as well. Few other aeronautical engineers in history have so transformed the world we live in.

Born in Seattle in March 1921, Sutter grew up in an era of open-cockpit biplanes, dashing aviators, and headlines proclaiming new aerial records or innovations seemingly every day. In love with it all, he built model airplanes, sketched imaginative flying machines, and plunked down 25 cents of his hard-earned paper route money whenever an airplane graced the cover of *Popular Mechanics*.

Joe was six years old when Lindbergh electrified the world by flying nonstop from New York to Paris. That audacious feat planted in the boy's mind the thrilling idea of a giant "aerial liner" that could take off and fly passengers in comfort to distant continents. It sounded like pure science fiction back then, but the thought captivated his daydreams and infused his sketches.

Sutter's boyhood home overlooked a little airplane company called Boeing. After delivering his newspapers, Joe would bicycle there to watch as new airplanes were rolled out and tested. He wanted to be part of aviation but didn't know how. The answer came at age 11 as he watched the sleek, all-metal Boeing 247 take flight in February 1933. He wanted to *design* flying machines. With this goal in mind, he applied himself to math and physics as his schooling progressed.

In the fall of 1939, as World War II broke out in Europe, Sutter took up aeronautical engineering studies at the University of Washington. He also underwent U.S. Naval Reserve officer training and worked summer jobs at Boeing. The United States too being at war by the time he graduated in 1943, he served aboard a destroyer-escort in the Atlantic Fleet before retraining and deploying to the Pacific Theater as an aviation maintenance officer.

Starting at Boeing in 1946, Sutter worked on the Model 377 Stratocruiser—the company's last propeller airliner—before getting in on the ground floor in jets. He rose quickly, assuming progressively greater responsibilities in the 707, 727, and 737 programs. Then in 1965, at age 44, the unexpected opportunity came his way to lead the development of the 747, a far bigger jet transport than any in service.

Airlines needed the 747 to carry two and a half times as many passengers as any other jet. To meet this requirement, Sutter and his team abandoned the expected double-decker configuration. Instead they laid out a cabin nearly 20 feet (6 m) wide and gave it two aisles.

The widebody jetliner had just come into being. This wide fuselage would also accommodate two side-by-side columns of cargo containers, making the 747 a great freighter when operated in all-cargo configuration.

The 747's development was truly one of the most colorful and precarious in history. Launched when the company was overextended, it nearly bankrupted Boeing. On the engineering front, Joe Sutter never had enough people, time, or resources to do the job even as his team contended with challenges as outsized as the airplane itself. Yet somehow he and the "Incredibles" overcame them all to successfully usher into revenue service one of the greatest airliners in history.

Joe Sutter retired in 1986 as Boeing Commercial Airplanes' executive vice president of engineering, but he always remained active as a consultant to Boeing. He was admired for the vision, integrity, and courage he brought to the 747 program, making it a textbook example of how to properly manage a major international industrial collaboration.

Soft-spoken and a good listener, Joe Sutter was known for his wisdom, intellect, and lack of evident ego. He won many honors over the years, among them the Guggenheim medal, Wright Brothers Memorial Trophy, National Medal of Technology and Innovation, and the Museum of Flight's Pathfinder Award. He served as a Museum of Flight board member from 1990 to 2005 and subsequently as a trustee emeritus.

To the delight of aviation buffs, Sutter participated countless times over the years in aviation-history events staged at the Museum of Flight and its William M. Allen Theater. Less well known is the fact that museum personnel often opened RA001, the first 747, for him to visit and show off to others. For Sutter, it was like visiting a very dear old friend.

Joe Sutter is interviewed in the main cabin of his old friend, the RA001.

Left: As the museum reached its 50th year, it celebrated the 34th annual Pathfinder Awards in 2015. Peter Morton is on stage with Pathfinder Alan Mulally.

Right: From the Red Barn to Raisbeck Aviation High School, Bill Boeing Jr. (center) was instrumental in the museum's first 50 years of growth.

CHAPTER NINETEEN

The Big Five-0

The new year 2015 began with the sad news that longtime museum benefactor William E. Boeing Jr. had died at his home in Seattle on January 8. Son of the Boeing Company founder, he took his first airplane ride in a Boeing Model 40 mail plane at age 5. He was with his father at the opening of Boeing Field in 1928 and continued his interest in aviation throughout his life.

Boeing Jr. had a long and successful business career owning helicopter companies, running a broadcast network, and developing commercial real estate, among other ventures. He also served on corporate boards. He was a committed philanthropist and a generous donor to the University of Washington, Seattle Children's Hospital, and the Museum of Flight, among many other organizations.

His support for the museum began early in the formation of PNAHF. He was instrumental in helping save historic building 1.05, the Red Barn, and donated funds for its move to Boeing Field. Over the years he made additional gifts to the museum, often stepping in at a crucial time to fund an acquisition or build an expansion, always with an emphasis on education. His financial and personal support was among the foremost of all contributors to the museum. The Board of Trustees voted to dedicate the 50th anniversary of the foundation to William E. Boeing Jr.

The museum gets and, in turn, gives support. An example of the foundation's generosity was evident at the Congressional Gold Medal Award Ceremony for the American Fighter Aces. The American Fighter Aces Association (AFAA) has official headquarters at the Museum of Flight. Their personal memorabilia are in the archives, and many of their stories are revealed in the J. Elroy McCaw Personal Courage Wing.

Left: Fighter Ace Commander Clarence "Spike" Borley, about to board a jet at Boeing Field to attend the Gold Medal award ceremony in Washington, D.C., in 2015.

Right: Space Squatch, or "Sasquatchtronaut," makes it back to the Great Gallery in time for the 50th-anniversary celebrations.

The Congressional Gold Medal, the highest honor Congress can bestow, was awarded in May 2014 to this small and aging group. In response to this national recognition, museum supporters pitched in to organize a formal presentation in Washington, D.C., a year later. At the time there were 77 remaining Aces, the oldest 104 and the youngest 72. This event was an opportunity to shine the spotlight on these warriors. "If there's an elite among fighter pilots, it's these men," commented AF Lt. General Charles "Chick" Cleveland, president of the AFAA.

Although some of the Aces traveled on commercial airlines to Washington, D.C., the museum enlisted 20 pilots and planes to fly 38 Aces and their families to the event. Among the fleet were five aircraft that MOF trustees made available. The presentation was held on May 20, 2015, in the Emancipation Hall at the Capitol Visitor Center. Hundreds of congressional, aerospace, and military leaders attended.

The MOF did what it could for a meaningful cause and the Aces. Everyone said that it was a great privilege to participate in both the presentation of the high honor and the reception that followed.

The momentum of planning was in full force in anticipation of the museum's 50th-anniversary gala. Building on the creative media campaign "Now everyone can be an astronaut" in early June 2015, the museum forewarned of an "Astronaut Invasion" attacking Puget Sound.

As a public art project, local artists painted and decorated 25 life-size fiberglass statues of astronauts that popped up around the city at museums, businesses, and tourist attractions and piqued the curiosity of residents and visitors. The unique and whimsical designs of "Astronauts on the Town" included "Space Tourist," carrying a camera and wearing a flowered shirt, shorts, and sandals; "Space Squatch," modeled on the mythical Pacific Northwest legend Sasquatch; and "Headin' Out," sporting a top hat for attending the 50th-anniversary gala. The six-foot statues traveled back to the museum for public display at the 50th-birthday celebration in September.

During the statues' exploits about town, the museum encouraged photo selfies and created a contest on Instagram at #astronautsonthetown, with all postings eligible for a drawing of two tickets on Alaska Airlines. Auctioned with silent bids online, the adopted astronauts went to new homes on terra firma.

The 50th-anniversary celebration on June 27, 2015, was a grand and extraordinary occasion. Valet attendants met the line of automobiles on the circular drive on East Marginal Way, and more than 800 guests in formal attire strolled the red carpet path up the steps to the entryway. The music of the Harry James Orchestra, directed by Fred Radke, introduced the festive mood, which was to continue for the next six hours. Once guests were inside the museum reception lobby, volunteers provided ticket packets that contained an elegant printed gala program, table assignment, and instructions for the evening. Nothing was left to chance. If any museum can throw a party, it is the Museum of Flight, and those who attend have high expectations. The festive, colorful evening, with reminiscences, tributes, and entertainment, exceeded the loftiest expectations.

Hors d'oeuvres during cocktail hour were selected and named for museum pioneers. Bill Allen's tribute offered a prosciutto-wrapped asparagus spear, and Bill Boeing Jr.'s a spicy fried chicken waffle cone. The Dick Bangert deviled eggs were a favorite. Bob "Swage" Richardson, Mr. B-17, was feted with a forest mushroom rosemary strudel that he might have found a bit fanciful. For those who knew Paul Whittier, the vodka and gin gimlets were masterful. Paul considered that a real gimlet was always made with gin, but he would have been proud and amused.

The entrance to the 50th-anniversary gala, flanked by "Astronauts on the Town" statues in gold livery.

Enjoying drinks in the Great Gallery, guests were immersed in the beauty of the spaces, the dramatic panoply of air and space objects suspended and pedestal-mounted within the gracious setting. Though the vast clear-span space with its resilient ceiling structure is massive, it is brought down to human level with the display of artifacts and well-situated interpretive panels. What could easily have been an intimidating and cold space is inviting and warm. It is the finest cocktail lounge and one of the great galleries of air and space in the world.

Those who have matured along with the museum over the years were impressed with its handsome maintenance. Carpets have been replaced, walls and trim repainted, lighting upgraded, and the color palette honored in all of the additions and remodels. The space seems as new as on the night it was christened in 1987.

At the appointed time, Fred Radke and his orchestra led the crowd to the entertainment tent on the east parking lot in what amounted to a New Orleans-style parade of music. Serving once again as emcee, Steve Pool, a longtime museum board member, welcomed the crowd and did his best to get the growing mass of energy seated. In his smooth and sincere style, Pool explained what the evening meant to him, as someone fascinated with flight since boyhood. He introduced the co-chairs for the evening, Bruce McCaw and Mike Hallman, who in turn paid their respects to the honorary chair, June Boeing, widow of the recently deceased William E. Boeing Jr.

50th-anniversary gala co-chairs Bruce McCaw (left), June Boeing, and Mike Hallman get the evening started.

Typical of so many museum events, the proceedings included a salute to military service with the announcement that, along with a host of dignitaries, there were two members of the famed Navy Seal Team 6 in attendance as well as Medal of Honor recipient and museum Pathfinder Joe Jackson.

McCormick & Schmick's, the museum's caterer and food service operator, prepared the superb dinner. With respect to the historical theme of the evening, the dining courses began appropriately with the T. A. Wilson-designated artisan cheese plate, featuring fine cheeses with fig and sour cherry jams, definitely a step up from the many so-called wine and cheese socials held in the Red Barn during the T. Wilson fundraising era. The starter salad was an Ibsen Nelsen baby romaine and Lola Rosa greens mix with pomegranate vinaigrette that the architect of the Red Barn and Great Gallery, an inveterate gardener, would surely have savored.

The intermezzo course was a palate-cleanser T. Evans Wyckoff peach with brut Champagne. The main course offered the Elliott Merrill miso-marinated Alaskan halibut with white cauliflower puree and a Jack Leffler captain-sized pan-seared filet of beef with roasted fingerling potatoes. Elliott, the first chair of PNAHF, and Captain Jack would have been proud to enjoy this kind of festive dinner in the museum they dreamed of building. Dessert was a beautiful medley of strawberry shortcake, sponge cake with fruits and chocolate, a double chocolate cake hazelnut brittle, and *panna cotta* with caramel corn labeled for William E. Boeing Jr. After this feast, servers offered Ned Skinner concert popcorn boxes in honor of a community leader who pitched in during initial funding. It was a fine dinner at a grand gala and a salute to some of the museum's essential contributors.

To follow was a video tribute to William E. Boeing Jr. His support of the museum spanned the entire 50 years, as he repeatedly stepped forward to help fund each stage of building. The animated crowd fell quiet, and with his widow, June, in attendance, the presentation concluded with the loudest applause of the evening. Particularly memorable was a moving rendition of "Over the Rainbow" with Fred Radke on trumpet and vocal by Brenna Whitaker—the perfect homage to a man who was known as the "patriarch of the Museum of Flight family."

Taking the stage as dinner lapsed into dessert, auctioneer Fred Northup Jr. steered the crowd through a quick and enormously successful live auction. Immense screens adorned all sides of the vast air-conditioned tent and offered continual visuals of people, aircraft, and events of the museum's first 50 years, providing a fascinating backdrop to the proceedings.

A few of the live-auction objects sold for well beyond the asking price, as bidders were caught up in the occasion and the atmosphere.

One of the items was an insurance cover with the signatures of Neil Armstrong, Buzz Aldrin, and Michael Collins—the three Apollo 11 astronauts. The signed "postal cover" stayed on Earth as insurance for the astronauts' families in the event of a catastrophe during the July 1969 mission to the Moon. Representing the Neil Armstrong Family Estate, Neil's son Mark donated the envelope to the museum for auction, and took the stage to explain its significance. With that introduction, the frenzied bidding began and its price climbed to $150,000, much more than its estimated value of $20,000.

Alaska Airlines and Holland America Line teamed to offer a seven-day package that included a cruise and first-class tickets for two passengers. Stein Kruse, CEO of Holland America, was at the gala and upped the offering to a 12-day cruise anywhere in the world. After the package sold for $12,000, Kruse, who was enjoying the evening, offered two more packages for another $12,000 each.

David Foster, the evening's entertainment director, donated the priceless experience of "hanging" with David for two days in Los Angeles and attending the rehearsal and taping of the PBS *Great Performances* concert featuring tenor Andrea Bocelli. The package for two included first-class flights on Alaska Airlines, with accommodations at the Beverly Wilshire. When the final bid hit $35,000, David announced another doubling of the package and $70,000 immediately flew into the museum's coffers.

With this quick auction complete, Fred ran a simple "Raise the Paddle" donation request, and in the prime category of $100,000, he received four immediate responses. With the targets for the evening met, patrons continued to make direct contributions at lesser levels that added up to more than was expected. The goal of $1.5 million was exceeded. The auction ran late, yet there were no objections from the enthusiastic crowd as no one wanted this memorable evening to end.

Taking the stage, musical icon David Foster stoked the crowd with his incredible show featuring international talent with local connections. Foster, a songwriter, composer, arranger, producer, and

Master of Ceremonies Steve Pool takes the stage at the 50th-anniversary gala dinner.

recording artist, was the perfect host for the evening. A friend of Bruce McCaw and Joe Clark, the gifted musician is also a pilot and aviation enthusiast. Foster was generous with his time, and the talent, personal resources, and connections he brought to the event. He commanded the stage as someone with 16 Grammys, an Emmy, and three Oscar nominations would do.

First up was Kenny G, a local boy with global performance credentials, notably one of the best-selling artists of all time. A flight enthusiast, he pilots a de Havilland Beaver floatplane. Kenny played his unique saxophone, winding through the audience and ending with one of his legendary long notes. Others who took the stage and earned the admiration of the large crowd included local talent violinist Anna-Ruth Boyce, playing a majestic piece, and rising star tenor Marcus Shelton, who astounded with his operatic range.

A highlight of the evening was the talented local EriAm Sisters, three young ladies who sang a distinctive blend of rhythm and blues. These sisters, though young, had been featured on the television program *America's Got Talent* and had performed locally for years. It was inspiring to learn that two of the three are proud graduates of Raisbeck Aviation High School.

Although not on the slate of entertainment, comedian Sinbad jumped on the stage to pepper the evening with jokes related to his

The Grand Finale, "I Believe I Can Fly."

experiences in Seattle and in flying. The incredible evening of entertainment was crowned when Kenny "Babyface" Edmonds took the microphone. This celebrated entertainer, singer, and songwriter worked effortlessly with David and the band in rounding out a fantastic show with some of his favorite tunes, even writing and singing an extemporaneous piece on the spot.

Along the line, members of the audience who wanted to sing took the stage, which was complementary to the joy of the evening. All the performers, along with Seattle's Total Experience Gospel Choir, joined in singing R. Kelly's Grammy-winning hit, "I Believe I Can Fly," bringing the evening to a stirring finale.

The show was the longest in museum history, yet deservedly so, as it was a 50th-anniversary party not to be forgotten. For the gala program, Charles and Lisa Simonyi wrote a perfect tribute: "In its first 50 years, the Museum of Flight has told the stories of flight and space, celebrated the heroes who have taken to the skies and beyond, and inspired those who have dared to push the envelope of exploration. We can't wait to see what the next 50 holds."

Pathfinders

In early October 2015, a tux-optional crowd of 600 arrived for the annual Pathfinder banquet. This 34th edition honored astronaut Michael P. Anderson in the "at-large" category and Elling Halvorson and Alan Mulally in "operations." Beyond the accomplishments of these recipients, the evening was a perfect expression of the integrative excellence of the museum.

Initiated in 1982, Pathfinders is an innovative form of an aerospace hall of fame. It was and is a joint venture of the museum and the regional chapter of the American Institute of Aeronautics and Astronautics. The idea, spawned even before there was a complete first phase of the museum, was to recognize Pacific Northwest leaders in the categories of aerospace, flying, manufacturing, engineering, operations, education, and an at-large category.

In that first year, inductees included William Boeing Sr. in the manufacturing category, former Boeing CEO Clairmont L. Egtvedt at-large, Thomas F. Hamilton and Louis S. Marsh in engineering, famous test pilot Leslie R. Tower and trans-Pacific record setter Clyde E. Pangborn for flying, and bush pilot Noel Wien for operations. It was a signal group that set a high standard. These recipients, called Pathfinders for their leadership qualities, would, in turn, enrich the museum stories of individual accomplishment and serve as motivation for students. The idea worked well, from the first gathering in a hotel ballroom through the years to the annual gala it has become.

In attendance on this evening in 2015, former director Lovering and longtime museum board member Brien Wygle shared recollections of the program's inception. Wygle, later a Pathfinder recipient, represented AIAA in that first year, 1982, and went on to lead the program for years. He expressed pride in what the program and the annual event have come to mean to the museum.

The 34th annual event was a festive and educational experience held in the Great Gallery with tables set under the Blackbird, which seemed an appropriate recognition of the leadership exploits of those honored. A large crowd assembled for the evening, including many of the previously recognized Pathfinders and a goodly collection of longtime board members, supporters, and donors. The occasion is representative of the importance of these events to the museum and its educational mission.

Top: Past Pathfinder recipients are recognized at the annual banquet.

Bottom: Museum CEO Doug King presents Michael P. Anderson's posthumous Pathfinder Award to his family.

The evening's three recipients were all individuals of great accomplishment. Lt. Col. Michael P. Anderson, astronaut, was a young man from a modest family in Spokane, Washington, who set his goal to be an astronaut after reading stories of space exploration. A motivated student, he worked hard in school, graduating with honors from the University of Washington in physics and astronomy. Anderson went on to a decorated career in the Air Force and was selected into the NASA astronaut corps in 1994, just as dreamed. He was on the crew of two space missions on the shuttle. As payload commander on the *Columbia*, he perished with the rest in the re-entry accident in 2003. His story was inspirational to all in attendance and poignant for his family, who participated in the tribute.

Elling Halvorson (left) receives his Pathfinder Award from Doug King.

Elling Halvorson, recognized in the operations category, is a legendary entrepreneur who has advanced vertical lift in construction, engineering, travel, and tourism. After graduating from Willamette University in Oregon with a degree in economics, he joined his family's construction business. Purchasing his first helicopter, a Bell 47-G3B1, he hauled workers and materials for a project to install a microwave tower on top of a peak in the Sierra Mountains. One of Halvorson's significant achievements was the building of a 13.5-mile water pipeline in the Grand Canyon using a fleet of helicopters. This three-year effort remains the largest helicopter-construction project to date. His work in the Grand Canyon stimulated his ventures in helicopter sightseeing excursions, where he was instrumental in developing technologies to enhance the tourist experience.

Alan Mulally was inducted for his signal career in operations, serving as a leader in aerospace as an executive at The Boeing Company, and finally as the celebrated turnaround CEO of the Ford Motor Company. Mulally graduated with an MS in aerospace engineering from the University of Kansas. Working for Boeing, he served on all of the airplane programs from the 727 to the 787 and won industrial awards for his contributions in aerospace. In 2006, he joined Ford Motor Company as CEO, initiating his teamwork concepts into the brand "One Ford." His work put the company back on track for financial stability and technological leadership. For these efforts, *Time* magazine recognized him in 2009 as "one of the world's most influential people."

These recipients or their representatives had many stories to tell, inspiring to all but particularly to the young people in the audience. The students of Raisbeck Aviation High School consider the Pathfinder banquet their "homecoming," as they don't have a full sports program. This evening incorporates their focus on aerospace technology and history with a special occasion in the school year.

Students spend time researching the stories of the recipients, and they volunteer and assist in preparing for the event. They are coached on table etiquette for a gala dinner with lots of flatware and plates. The students dress in formal attire and are expected to circulate and talk to others in attendance. On this evening, a string ensemble from the school played background music.

It is part of their training to express what they have learned and to communicate with those in the profession. They do an impressive job, and their participation has provided another level of significance for this traditional annual event.

The Pathfinder program is so much more than an ordinary Hall of Fame. It has a scale in the way it incorporates museum events and programming, ultimately providing the essential stories of human achievement that enrich all museum exhibits. It is a tradition with solid roots, reaching out to recognize those who lead the way. It brings their stories into the museum, gives them life and meaning, and emphasizes the individual nature of aerospace advances. Perhaps most important, the program serves as a motivator for young people, particularly the students at Raisbeck Aviation High School.

Peter Morton (left) interviews 2015 Pathfinder recipient Alan Mulally during the evening's program.

The museum's Boeing 80A was a perfect backdrop as students launched paper airplanes in celebration of the museum's largest private contribution to date.

On July 23, 2015, the Museum of Flight announced the largest private contribution in its history. The ceremony was a colorful and jubilant affair staged in front of the Boeing 80A in the T. A. Wilson Great Gallery. The huge gift of $30 million, with $15 million equally from The Boeing Company and Bill and June Boeing, was announced and pledged to the new Boeing STEM Academy in support of education.

Media representatives joined an enthusiastic audience of museum leadership, elected officials, educators, and hundreds of students from various museum programs, including recent graduates of Raisbeck Aviation High School. Appropriately, two alumni of the Washington Aerospace Scholars program introduced the announcement. Students wearing the colors of their education programs contributed to the fun and energy of the occasion. The ceremony concluded with a mass launching of blue paper airplanes into the expanse of the Great Gallery.

Ray Conner, CEO of Boeing Commercial Airplanes, took the dais to explain that this timely gift would focus on the tools necessary to prepare young people for jobs in the expanding tech economy. "Jobs related to science, technology, engineering, and math represent the future. But thousands of openings will go unfilled because Washington State doesn't have enough qualified candidates to fill them. We are confident the Museum of Flight is the perfect partner to help us expand the pipeline of diverse, talented young STEM professionals in Washington and beyond. The young people who will benefit from the Boeing Academy for STEM Learning represent the future of our community."

June Boeing, widow of Bill Boeing Jr., spoke briefly and humbly on behalf of the couple's matching gift of $15 million. "It is such an honor to announce today that Bill, in one of his last acts of generosity before he passed away in January, agreed to join in partnership with The Boeing Company to also invest $15 million personally in the museum and its mission. Thank you for the opportunity to partner with such a great company to support the educational efforts of such a world-class institution."

Speaking for the museum, CEO Doug King welcomed the gift as "transformational." He noted the museum's 50 years of commitment to education, underscoring this day and this generous gift as a resource to take a significant step forward. "This year marks five decades of providing educational programs to students of all ages. With this combined $30 million investment, we've been launched on our way toward transforming our museum, our educational mission, and the community we serve. You have truly made a gift to the community through the museum, and validated our vision to be the foremost educational air and space museum in the world!"

Ray Conner set high expectations for the Academy, stating, "This gift will allow the Museum of Flight

to significantly expand STEM opportunities from kindergarten through college. The number of students will double by 2017, and double again by 2019." This objective signified a worthy return on investment—to serve 5,000 students by the end of the current decade in expanded educational programs—and challenged the Academy to recruit half that total from the underserved—women, minorities, and other disadvantaged students. The museum was already well down the runway in these areas with the Michael P. Anderson Memorial Aerospace Program for underserved children, introducing them to the museum programs and connecting them with volunteer mentors within the aerospace industry. Amelia's Aero Club attracts middle-school girls to STEM activities.

Top: June Boeing and Boeing Commercial CEO Ray Conner represented the Boeing family and company for the combined gift.

Bottom: Numerous STEM education programs will continue to benefit from the generous gift from the Boeing Company and Bill and June Boeing.

Top: Visitors during
the 50th-anniversary
party, through
the wings of the
Wright Flyer.

Right: Astronaut on
the Town "Destiny,"
by Camille Patha.

The additional key educational components of the Academy are the existing successful programs—Washington Aerospace Scholars, Aerospace Camp Experience, the Hallman Spaceflight Academy, Challenger Learning Center, and the Aviation Learning Center—which will improve and expand as motivational and educational resources, among the best of their kind in the nation.

Also benefiting from this generous gift is one of the museum's newest educational programs, Private Pilot Ground School. This is a bridge program from early-learning initiatives into practical workforce training and serves as a model for future technical programs.

The experienced and talented education department, which has expanded over the last decade, uses the extensive facilities and priceless collection to offer what is equivalent to a year of enhanced education. Their endeavors include Teacher Professional Development, providing training in all of these initiatives with the objective of strengthening service to schools across the state.

The positioning of the presentation platform was ideal. Behind the presenters and graphic visuals of the Academy loomed the Boeing 80A, the artifact that helped to forge the museum foundation 50 years earlier. This occasion was more than another photo opportunity and media release for a major gift. It was a time to reflect upon the museum's modest beginnings and appreciate this day, this support, and this departure point—a wrapping-up of the first 50 years and a spectacular blastoff for the next 50. Bruce McCaw summed up the celebration, saying, "This is clearly the future, and the pieces are in place."

On September 19, 2015, the museum hosted its 50th-anniversary party for members and the public. Entry lines snaked out the doors and along the walkways for much of the afternoon, with 5,000 visitors joining the festivities. The theme celebrated the decade of the museum's incorporation with a 50-cent admission and a sixties fashion contest. Games, a scavenger hunt, and a design-your-own astronaut coloring project rounded out the day, complete with McCormick & Schmick's birth-day cupcakes. All 25 of the "Astronauts on the Town" convened under the Blackbird, where the museum announced the winner of the Instagram selfie contest.

The birthday party took on a poignant significance to those supporters involved long enough to experience the journey, as they realized just how far the Museum of Flight had come in accomplishing its dream.

Anne Simpson became the first female board chair of the Museum of Flight in 2016.

Douglas R. King
President and CEO, 2010-

CEO Doug King (left) with trustee Bill Ayer in front of the museum's 727.

My first contact with the Museum of Flight was in 1991, when I was working for the Challenger Center for Space Science Education. The museum was greatly expanding its role in education, and the Challenger Center was looking for partners for its network of Challenger Learning Centers. I met several times with Howard Lovering, and he had a bold vision for the museum to do something very different: a dedicated field trip destination to attract and train teachers and students from across the region. The result was a facility that over the last 23 years has taken literally hundreds of thousands of local students "to space" and inspired them about STEM studies and careers in aerospace.

As I continued to visit the museum over the ensuing years, it expanded dramatically, and its long-term commitment to become a leader in STEM education was obvious. Its vision statement became *"To be the foremost educational air and space museum in the world."* The growth of its own programs and reputation matched its impressive physical, collection, and exhibit progress. By the time I was interviewing here in 2010, there were even plans to build a high school on the campus.

All this change was accomplished during a turbulent decade for museums across the country. The aftermath of 9/11, the dot-com boom and bust, and the national financial debacle of 2008 had debilitating effects on us all. But the leadership of the Museum of Flight had persevered. A series of strong chairmen and committed trustees supported Ralph Bufano and Bonnie Dunbar to build the J. Elroy McCaw Personal Courage Wing, the T. Evans Wyckoff Memorial Bridge, the "Rendezvous in Space" exhibit, and the "Space Gallery," and to fill them with amazing new artifacts and exhibits. Planning for an additional new gallery to tell the story of the space shuttle era and the future of space was under way. And they had also weathered the financial storm and, with the incredible involvement of Bill Boeing Jr., had the balance sheet healthy enough to be able to look forward. This process and progress had not been easy. It had taken dedicated effort from many trustees, led by Bruce McCaw, Kevin Callaghan, Gene McBrayer, and Mike Hallman.

Mike Hallman served as volunteer Interim President in 2010 and became Chairman in 2011. When I came aboard he said it was time to update our long-term strategic plan, but he put it in much more elegant terms: "Uniting the Board about the future is going to take an effort that gets them all to climb up a tree together, look off into the distance, and decide what mountain we are headed for!" This process, led by Mike and Gene McBrayer, involved not just the Board, but the staff, docents, volunteers, and partners in the community over most of 2011. The result was "Vision 2020," which has guided our progress since. The plan was simple, really, calling for mutually supportive world-class aspirations in our collection, visitor experience, and education efforts, and a strong financial plan to sustain their growth.

Mike also recruited trustee Bill Ayer to be vice chair, and to help us "climb down out of the tree and figure out what path to take." As chairman, Bill led us to a very pragmatic rolling three-year planning horizon and a comprehensive fundraising strategy that has guided us since. The results have been dramatic growth in the size of the museum, the breadth of our visitor experience, the scope of our collecting, restoration, and exhibit efforts, the depth of impact of our education efforts, and the upgrading of existing facilities, as well as a stronger financial base. We opened the Charles Simonyi Space Gallery and Raisbeck Aviation High School, which is now ranked as the top public high school in the state by *U.S. News & World Report*. We plan to complete the *Inspiration Begins Here!* Campaign this year. Then, under the leadership of new Chairwoman Anne Simpson, we will "climb up the tree again" and update the long-term plan.

It has been very gratifying to be a part of this. I have been a fan of the Museum of Flight for almost 25 years. But what I have come to learn in the last six is that while our planes and artifacts and programs are amazing, it is the people of the museum that make it such a special place. The trustees, committee members, docents, and volunteers—all of whom donate their time, expertise, and resources out of love for the mission and those we serve—are an incredibly driving and motivating force. Some are new to the museum, and others have been with us for many years. Seven former chairmen still serve actively on the board. Emeritus Trustee Bob Mucklestone was even there when the Pacific Northwest Aviation Historical Foundation was chartered, and is still an active leader today. They are a truly inspirational group.

They are complemented by the most dedicated, committed, and professional staff that I have ever served with, in either the nonprofit or for-profit world. Both groups take their duty to serve the community very seriously and have created a very strong culture of mutual respect and service. It is an honor to work with them.

With this great team in place, I look forward to completing the first phase of Vision 2020, to recharting the course beyond, and to the great success of the Museum of Flight in its second 50 years.

— Doug King

Left: The Saturn V rocket rolling out of the Vehicle Assembly Building at Cape Canaveral on its massive crawling gantry.

Right: Apollo F-1 thrust chamber embedded in the ocean floor, three miles beneath the surface of the Atlantic Ocean.

CHAPTER TWENTY

To the Horizon

In 2013, Amazon founder Jeff Bezos was looking to the skies to develop consumer suborbital space travel with his company Blue Origin, but he was also searching underwater in hopes of retrieving some Apollo parts from deep in the Atlantic Ocean.

Bezos Expeditions was exploring the ocean floor for the Rocketdyne F-1 engines that launched the Apollo 12 and 16 missions to the Moon. These engines had lifted the Saturn V rockets 40 miles up to the edge of space, where they detached from the capsule and plunged back to Earth. Somewhere off the coast of Cape Canaveral in Florida, they had descended 14,000 feet to the bottom of the Atlantic, where they rested for the next 44 years. The difficult submarine mission was challenging, but with the modern aids of deep-sea sonar and robotic technologies, Bezos Expeditions' skilled crew of 60 found and retrieved the remains of the massive engines.

Bezos was one of those youngsters who was thrilled and motivated by the early space program. He was fascinated at watching Neil Armstrong walk on the Moon, and said that he understood what it meant for "this idea of science, engineering, technology, and exploration."

On November 19, 2015, the 46th anniversary of the Apollo 12 moon landing, NASA and Bezos Expeditions announced that the restored F-1 engine components would be on permanent display at the Museum of Flight as an inspiration to a new generation.

"To bring those pieces up on deck and actually touch them, that brought back for me all those feelings I had when I was five years old and watched those missions go to the Moon," Bezos said at the ceremony. "If this results in one young explorer, one young adventurer, one young inventor, doing something amazing that helps the world, I'm totally fulfilled."

Top: Saturn V first-stage booster with F-1 engines being assembled at NASA's Michoud Assembly Facility in New Orleans.

Bottom: Jeff Bezos tests communication systems before the first flight of the New Shepard space vehicle.

The idea of a space component in the Museum of Flight was slow to rise but did take off at the opening of the Great Gallery with the participation of the Mercury 7 astronauts. With the continued growth of the space themes, a coterie of astronauts and cosmonauts and private space explorers found the museum a comfortable place for memories and programs. Supporter Pete Conrad, the mission commander on Apollo 12, left some of his valuable personal items to the museum, and it was appropriate that the retrieval and gift of the F-1 engine commemorated his momentous flight.

These extraordinary donations of the Apollo engines were made just days before Bezos' private company Blue Origin was scheduled to test its New Shepard propulsion module, named after astronaut Alan Shepard. On November 23, 2015, the New Shepard capsule and its BE-3 booster engine blasted from a test site in West Texas and thundered to the boundaries of space. At an altitude of more than 300,000 feet, the capsule and booster separated, with the engine dropping tail first as planned. The millions of people who watched this test online marveled at the recorded images: the rocket plunging down fast, braking quickly at a low altitude of 5,000 feet, and with the BE-3 engine once again igniting to slow the return, the rocket settling down to a remarkably soft landing on its pad. When the dust cleared, it was evident that this was a perfect landing and proof of a reusable system with the Vertical Takeoff and Vertical Landing concept for this company and its spaceship. The New Shepard capsule, with its open parachutes, also landed safely at the test site.

This was the first time for such a demonstration, and it underscored the incredible advances that private ventures were making with new space travel initiatives. Puget Sound has been the hometown hotbed for some of these enterprises, boasting headquarters or offices for Vulcan Aerospace, Space Exploration Technologies, Planetary Resources, and Spaceflight, companies that are energizing the domestic space sector with technology transfers from the computer fields. This generation of space explorers and entrepreneurs is revolutionizing the national profile of science, technology, and business organizations, providing an unexpected era in aerospace that will trigger innovative exhibits surrounding the latest artifacts.

The Museum of Flight, from model and early rocketry to the Mercury and Apollo programs, the Shuttle era, and all of the associated programs and artifacts, encompasses a significant aerospace interpretation in its first 50 years of growth. Now, with the anticipation of additional space objects from others in these private initiatives, the aerospace story will continue to be revealed. That is how it is within the continuum of time, scientists, technologists, builders, and explorers.

How will the Museum of Flight grow and expand in its collection, facilities, programs, and public service? What is the vision? What are both the practical and imaginative thoughts for the long-term future?

Top: F-1 booster engine injector plate.

Left: The museum's high-fidelity mockup of the International Space Station's Destiny Laboratory module was acquired in 2006.

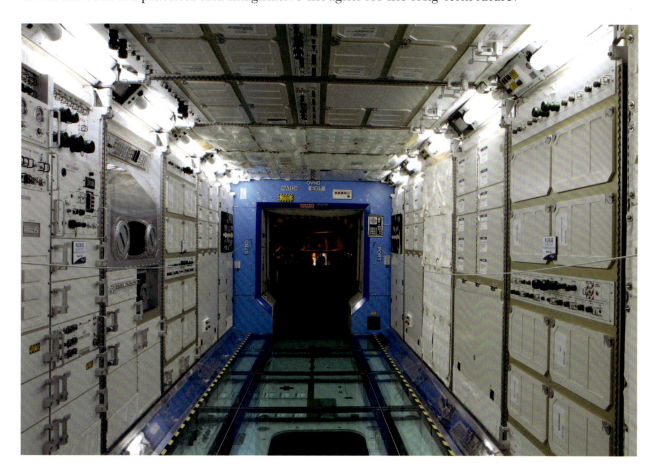

The Six

The Museum of Flight is a team of board, staff, volunteers, and supporters. Staff who work assignments each day also enrich the institution with ideas and skills and passion. Among those few who have worked continuously for more than 25 years are veterans who have engaged the institutional growth since its early phases. The employment of six seasoned pros ranges from 26 to 32 years each, and the performance of these troops combines for a total of 174 years and more than 350,000 hours of museum service.

Alison Bailey, hired on in 1984, is the youngest of the six but is the most senior in service. She personifies this group with her hard work, vast knowledge of the institution, and total dedication to the cause. Alison has worked in development from the early campaign for the Great Gallery through all that followed. She takes a lead role in events and galas at the museum. As Associate Director of Development, Alison knows thousands of members and donors and treats all as treasured friends. As public as she has to be, Alison also has a stealth gene and prefers not to take the spotlight.

Likewise, Rosemarie "Rosie" Gran, 1985, is the voice of the museum, the greeter at administration, and the go-to person for all who do business there. She began as a volunteer, worked in the gift shop, and jumped at the chance to woman the front desk at reception and ascended to the rank of the museum's first Telecommunications Manager. One cannot imagine anyone else in that position. She has a smile for every visitor, manages a calming influence during a flurry of telephone insinuations, and is always at the ready with coffee or one of her red licorice whips for the needy.

Christine Runte came on board in 1986, beginning her work with the campaign staff and then transferring to the new library and archives in the Great Gallery. Christine could not have imagined what she would witness in the collections business over her tenure as the museum's official Registrar. It has been an adventure that she continues to pursue with renewed interest every day.

Jeff Frignoca, class of 1987, was one of the first to implement a new kind of visitor-service mentality, rising to manager of this division, where he served for nearly 20 years. Switching gears in 2006, he became Product Development Manager, creating branded merchandise and making money for the gift shop. Jeff is an aviation historian, like so many others who have worked on staff. He was production manager with editing duties on both the *Personal Courage Wing* catalog and *The Museum of Flight: A Collected History*, and considers those the most important projects he has supported.

Eric Betten joined the staff in 1990, migrating from an audio engineering background employed in sound and studio recording. A guitarist he knew was working a day job as Security Manager at MOF and recruited Eric to the force. Eric has been at the museum since, also supporting the A/V department, modernizing fire and security systems, and training staff on safety procedures. In his words, "The chances are that if it's red and hangs on the wall, Eric takes care of it."

Fenton Kraft, 1990, is the new kid among the sensational six, but in his prowling the physical facilities he has come to know many secrets. Fenton oversees the museum's mechanical, electrical, and building automation systems and for 17 years also managed the museum's rapidly expanding computer network. He designed and installed improvements including a high-efficiency boiler plant, air-conditioning systems for educational and exhibit spaces, a humidification system to protect the artifacts in the Great Gallery, and air-conditioning systems in several aircraft, including the Air Force One 707 and the 787. Fenton, like the others, seems to have a perpetual smile and good humor, even while responding to the constant calls for fixes to the operating plant.

These six are just a portion of the working staff, but their continuous service brings a high level of skill honed by long experience. They are representative of the exceptional staff component of the museum team.

Top: The Simonyi Space Gallery, next to the newly opened Aviation Pavilion, 2016.

Bottom: The museum's 2016 restoration of the prototype Vought XF8U-1 Crusader.

The museum's prime objective—to maintain its preeminent position as the foremost educational air and space museum in the world—will take effort. This is a daunting task, and the challenge to upgrade and enhance programs and curricula has already been answered with the recent establishment of the Boeing Academy for STEM Learning, a big step into the future. While maintaining this leading position in education and sustaining operations, the museum will look for additional opportunities to grow.

Physical expansion was a constraint at the opening of the Great Gallery. The museum was entirely hemmed in within the original seven acres, and the words were "no more land is available, period." It did not take long for the museum to realize it had to expand.

Who could have foreseen the opportunities that would become available when, surprisingly, The Boeing Company moved its corporate headquarters to Chicago? Suddenly, land was available. The museum made purchases and trades of properties with Boeing, the airport, and private owners. These sites allowed for building the archives, Personal Courage Wing, Space Gallery, Aircraft Pavilion, and Aviation High School. The museum rocketed from its formerly landlocked site into a much-enlarged campus that bridged the arterial and doubled in size.

Growth and development can also be elsewhere. It is possible there will be branch locations, additional facilities at Paine Field, or partnerships with other, smaller regional flight attractions. The MOF has much to offer in shaping a global network. Programs such as Aviation Learning Center, Space Camp, the ACE program, hardware, software, and operating experience are plausible franchises. The intellectual capability and the operations experience are transferable to other locations and a means to expand public service.

Perhaps the future of this flight museum is in the cloud. Already active in online instruction and information, with robust social media, MOF will likely be a leader in the digital experience. So many opportunities are launching that it's hard to know where this fast-changing environment might take the entire museum world. The MOF must be prepared to take on new challenges, seize opportunities, and continue to build public service as a flight museum leader.

The Museum of Flight brand is known and respected far and wide. The skilled museum team of board, staff, and volunteers is a valuable resource with decades of operating experience. The foundation is adept at acquiring capital, collecting, and building, and will lead the museum into other successful ventures. The possibilities are endless.

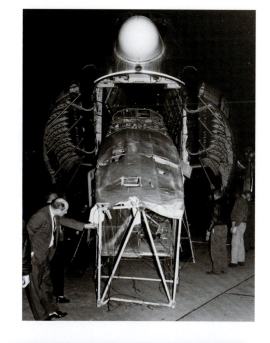

On February 25, 1964, enthusiastic trophy hunter Captain Jack Leffler was at McChord Air Force Base, waiting for a Douglas C-124 Globemaster. Winging its way south from Elmendorf Air Force Base in Alaska, the plane carried the precious remains of a broken-down 80A, rescued from an Anchorage landfill and later cataloged as the museum's first artifact, NC-224M. Jeff Bezos was little more than a month old when the 80A was heading to Seattle aboard that Globemaster.

Fifty years from now, in time for the museum's hundredth birthday, some individual might be moved to present the museum with his or her

era's equivalent of an 80A retrieved from a dump or an Apollo engine raised from the ocean floor. Until then and into the future, it is the museum's responsibility to both preserve the past and instill dreams of flight and space exploration:

For Future Generations.

Top: Unloading of the fuselage of the 80A from the C-124 Globemaster at McChord Air Force Base, 1964.

Bottom: The view from the cargo bay of the Space Shuttle FFT features a half-scale model of the Hubble Space Telescope.

Museum of Flight

Airplane Type	Pre-1910	1910-1920	1920-1930	1930-1940
Civilian			3 4	5 7 8 9 10 11 12 13 14 15 16
Commercial		25	27 28 29 31 32	33 35 36 37 38
Experimental	50 52 53 54 55			
Military		82 83 84 85 87 88 89 90 91 92 93 94 95 96 97 98 99 100 101		103 106 107
Space				
Helicopter				

Events

1963
Harl Brackin and Jack Leffler pursue the remains of a 1929 Boeing Model 80A in Anchorage, Alaska

1964
Harl Brackin develops a charter for what will be the Pacific Northwest Aviation Historical Foundation

1965
PNAHF founded and registered as a 501(c)(3) organization

1966
Renton Aviation Festival

1966
First annual membership meeting of the PNAHF (May)

1967
PNAHF performs fly-over at Seafair Gold Cup Races

1969
The Boeing Company makes its first large gift to PNAHF

1970
Boeing 247D flight to San Francisco for United Airlines celebration

| | **1965** | | **1970** | | **1975** |

Exhibits

1968
The City of Seattle leases a building near the Space Needle at Seattle Center to PNAHF for an aerospace exhibit.

1974
Boeing building 1.05, "The Red Barn," is listed on the National Register of Historic Places

Artifacts

1964
1929 Boeing 80A-1

1967
1934 Boeing 247D

1969
1951 Boeing B-47 Stratojet

1969
1948 Lockheed TV-1 Shooting Star

1969
1916 Boeing B&W replica (1966)

1968
1959 Northrop YF-5A Freedom Fighter

1969
1943 General Motors FM-2 Wildcat

1971
1929 Curtiss-Robertson Robin C-1

Board Chair

1965-1973
A. Elliott Merrill

1973-1974
Thomas R. Croson

1974
James Dilonardo

| | **1965** | | **1970** | | **1975** |

CEO

1965
Cofounder of PNAHF
Harl V. Brackin Jr.

1976
Yesterflight newsletter

1976
Red Barn Open House draws 20,000 people

1981
Boeing employees pledge $970,000 to museum

1980
Museum receives a $1 million gift from The Boeing Company

MUSEUM OF FLIGHT®

1982
Museum changes name to Museum of Flight

1982
First Pathfinder Awards Dinner

1984
Concorde "Flights to Nowhere"

1984
Douglas "Wrong Way" Corrigan

1986
MOF receives accreditation from the American Association of Museums

1985
The museum has over 168,000 visitors this year

1985
The Boeing Company celebrates the 50th anniversary of the B-17 Flying Fortress with 50,000 in attendance

1988
"Friendship One," a specially outfitted 747, takes off from MOF on an around-the-world speed record

1988
Museum holds its first air show, the "Emerald City Flight Festival"

1989
A British Airways Concorde returns to Seattle for a "Flight to Nowhere" fundraiser

1989
Museum celebrates the state centennial with "Wings Over Washington"

1990
Museum hosts 15th reunion of Apollo-Soyuz program

1980 · **1985** · **1990**

1975
Model of the Red Barn Air Park is presented and the building is barged to Boeing Field

1977
Education program under Georgia Franklin is initiated

1977
Seattle Center location is remodeled

1979
Museum board adopts building program for Red Barn and Great Gallery at Boeing Field

1983
The Red Barn opens to the public

1985
Groundbreaking for the Great Gallery

1987
Great Gallery Ribbon Cutting and Grand Opening

1989
"The Hangar" opens, to the delight of the museum's youngest visitors and national recognition

---

1977
1928 Boeing P-12 (Model 100)

1983
1917 Curtiss JN-4D Jenny reproduction (1982)

1987
1940 Douglas DC-3

1987
1968 Taylor Aerocar III

1990
1963 Boeing 727-022

1990
1969 Boeing 747-121 (RA001)

1985
1943 Boeing B-17F Flying Fortress

1982
1951 Boeing WB-47E Stratojet rollout

1986
1926 Ryan M-1

1989
1980 Lear Fan 2100

1974-1977
Robert S. Mucklestone

1977-1979
George C. Briggs

1979-1988
Richard E. Bangert

1980 · **1985** · **1990**

1977-1991
Executive Director
Howard C. Lovering

Collection of Major Artifacts

1940-1950	1950-1960	1960-1970	1970-1980	1980-Present

2006
Washington Aerospace
Scholars educational
program

2006
"Leonardo da Vinci:
Man, Inventor, Genius"
blockbuster exhibit
draws 61,000 visitors

Ieroes"
tes
vith Jim
I Haise,
rmstrong

2008
Gala Auction for
Education raises $1.7
million

2007
"Exploring the New
Frontier" gala honors
the X-15 program

2009
Gala Auction
honoring Apollo 11

2011
First Red Barn Heritage
Award presented to
William E. Boeing Jr.

2012
Largest gala to date,
with over 40 space
luminaries and
astronauts

2011
K-12 education reaches
an all-time high of
151,000 participants

2010
World War II Hangar
Dance Gala with the
B-17F

2015
"First Fifty Gala"
celebrates the
museum's history

2013
Attendance soars
to over 540,000
this year

2016
"Inspiration Begins Here!"
campaign surpasses $77
million goal

2016
The museum is
headquarters for The
Boeing Company's
100th-Anniversary
"Founders Day"
Celebration

2005 | **2010** | **2015**

---

ith
ool
tion High

earning
ening

w
age

elcomes
d in honor
IcCaw

2006
Opening of the
Kenneth H. Dahlberg
Military Aviation
Research Center

2007
The museum aquires
6.47 acres on the
west side of East
Marginal Way S.

2007
"Space: Exploring
the New Frontier"
exhibit opens to
the public

2008
T. Evans Wyckoff
Memorial Bridge
links the east and
west campuses

2009
Reaccreditation
from the American
Association of
Museums

2011
Launch of "Vision
2020," focused on
the next decade of
museum growth

2011
Charles Simonyi
Space Gallery
Grand Opening

2013
Raisbeck Aviation
High School
Grand Opening

2013
The William M.
Allen Theater
renovation is
completed

2014
Michael and Mary Kay
Hallman Spaceflight
Academy opens

2015
Construction begins
on the 3.2-acre
Aviation Pavilion

2015
The Alaska Airlines
Aerospace Education
Center opens

2015
The Boeing Company
and Bill Boeing Jr. gift
$30 million for STEM
learning

2016
Aviation Pavilion opens
to the public, housing
the Boeing 787, 747, Air
Force One, Concorde,
and the Boeing B-17,
B-29, and B-47 bombers

2007
1927 Boeing Model
40B replica (2000s)

2009
1954 Lockheed
Super Constellation

2010
1944 North American
P-51D Mustang

2011
1995 DG Flugzeugbau
Perlan Glider

2012
1959 Bell UH-1H
Iroquois "Huey"
helicopter

2012
2005 Blue Origin
Charon test vehicle

2012
1973 NASA Full
Fuselage Trainer

2013
1935 Lockheed
Electra 10-E

2014
2009 Boeing
787 Dreamliner

2015
Rocketdyne F-1
engines from
Apollo 12 and
Apollo 16

2005-2007
James T. Johnson

2007-2009
Robert J. Genise

2009-2011
J. Kevin Callaghan

2011-2014
Michael R. Hallman

2014-2016
William S. Ayer

2016-
Anne F. Simpson

2005 | **2010** | **2015**

2005-2010
President and CEO
Bonnie J. Dunbar, Ph.D.

2010
Interim President and CEO
Michael R. Hallman

2010-
President and CEO
Douglas R. King

1992
Museum welcomes the Blue Angels during Seafair

1992
MOF joins Boeing in hosting the 50th anniversary of the B-29 Superfortress

1993
Launch of the "Teaching Through Flight" educational resource manual

1990
MOF's "Wings Over the Pacific" exhibit opens in San Francisco

1995
The museum's first fundraiser, a USO-style Hangar Dance

1995
The final Flight Fest airshow

1996
"FlyerWorks" over Myrtle Edwards Park

1997
First fundraising auction gala raises $1 million

1998
The museum's B-17F takes to the skies after a 100,000-hour restoration

1999
MOF hosts its second major auction gala, "Out of This World for Kids," raising $2 million

2000
Education department grows to 14 full-time and 20 part-time educators in 18 core programs for K-12

2004
Membership reaches 22,000

2001
"Sky Without Limits" capital campaign for a new home for the Champlin Fighter Collection initiated

2001
Auction gala with Harrison Ford raises $2 million

2005
"Wings of gala celebr Apollo 13 Lovell, Fre and Neil A

1990 | **1995** | **2000**

1993
Multi-use Murdock Theater opens to the public

1992
The museum opens Challenger Learning Center

1995
The Hatfield Collection and the *Enola Gay* tapes added to the archives

1994
The Wings Café, Skyline Room, and state-of-the-art kitchen operated by McCormick & Schmick's

1994
Satellite store at Sea-Tac Airport opens

1997
New permanent exhibit "The Tower" opens to the public

1997
Elrey B. Jeppesen Collection aquired

2000
"Rendevous in Space: A Tribute to Pete Conrad" exhibit opens with gala event

2001
The Lear Archives are donated to the museum

2001
The "Flight Zone" replaces The Hangar

2002
Library and Archives move into their new home in the former Boeing 9-04 building

2003
Priceless collection of Wright brothers documents acquired

2004
Partnership v Highline Sch District's Av School

2005
Aviation Center op

2004
J. Elroy McC Personal Cou Wing Grand Opening

2004
New Lobby v visitors, nam of Keith W. M

1991
1963 Lockheed M-21 Blackbird acquired and entered restoration

1992
1945 Goodyear FG-1D Corsair restored

1994
1959 de Havilland D.H. 106 Comet 4C

1995
1971 Boeing Lunar Roving Vehicle

1995
1964 McDonnell F-4C (F-110A) Phantom

1995
1970 Grumman A-6E Intruder

1996
1958 Boeing VC-137B "Air Force One" SAM 970

1999
1934 Douglas DC-2

2003
1978 Concorde G-BOAG

2003
1967 Boeing 737-130

2003
Champlin Collection of Warbirds

1999-2001
Harold E. Carr

2002-2003
H. Eugene McBrayer

1988-1993
Robert E. Bateman

1993-1995
Harold F. Olsen

1995-1997
James A. Curtis

1997-1999
Mark E. Kirchner

2001-2002
David C. Wyman

2003-2005
Bruce R. McCaw

1990 | **1995** | **2000**

1992-2005
President and CEO
Ralph A. Bufano

Major Artifact List

1. Gas Balloon
2. Curtiss Pusher
3. Cessna CG-2
4. Curtiss-Robertson Robin C-1
5. Aeronca C-2
6. Beech 17 Staggerwing
7. Bowlus BA-100 Baby Albatross
8. Granville Brothers Gee Bee Z*
9. Heath Parasol
10. Howard DGA-15P
11. McAllister Yakima Clipper
12. Piper J3C-65 Cub
13. Stinson SR Reliant
14. Taylorcraft A
15. Taylorcraft BC-12D
16. Stinson Model "O"
17. Ercoupe 415-C
18. Fairchild 24W
19. LET LF-107 Lunak
20. Learjet 23
21. Lamson L-106 Alcor
22. Stephens Akro
23. DG505M Perlan Glider
24. Beech Starship 2000A
25. Boeing B&W*
26. Glasair S11-S FT
27. Alexander Eaglerock
28. Boeing 40B
29. Boeing 80A-1
31. Ryan M-1
32. Swallow Commercial
33. Boeing 247D
35. Douglas DC-2
36. Douglas DC-3
37. Stearman C-3B
38. Lockheed 10E Electra
39. Antonov An-2
40. Lockheed 1049G Constellation
41. de Havilland D.H.106 Comet 4C
42. Lockheed JetStar
44. Boeing 727-022
45. Boeing 737-130
46. Boeing 747-121
47. Aerospatiale/BAC Concorde
48. Boeing 727-223
49. Boeing 787 Dreamliner
50. Chanute-Herring 1896 Glider*
52. Leonardo da Vinci Il Siglo*
53. Lilienthal 1893 Glider*
54. Wright 1902 Glider*
55. Wright 1903 Flyer
56. Bensen B-8M Gyro-Copter
58. Sorrel Parasol
59. Taylor Aerocar III
60. Aerosport Scamp
61. Bowers Fly Baby (2)
62. Rutan VariViggen
63. Thorp T-18 Tiger
64. Bede BD-5B
65. Cascade Kasperwing 180B
66. Eipper Cumulus VB

67. Lear Fan 2100
68. Monett Monerai
69. Rotorway Scorpion Too
70. Rutan Quickie
71. Williams International X-Jet
72. Insitu Aerosonde
73. McCready Gossamer Albatross II
74. Pterodactyl Ascender
75. Pterodactyl Ascender II
76. Rotec Rally IIIB
77. Rutan Voyager**
78. Sather DEX-1 RPV
79. Wizard J-2
80. Task Silhouette
81. W.A.R. P-47 Thunderbolt
82. Albatross D.Va*
83. Aviatik D.I
84. Caproni Ca 20
85. Curtiss JN-4D Jenny
87. Fokker D.VII*
88. Fokker D.VIII*
89. Fokker Dr.I*
90. Fokker E.III*
91. Nieuport Type 24*
92. Nieuport Type 27*
93. Nieuport Type 28*
94. Pfalz D.XII
95. RAF S.E.5a*
96. Rumpler Taube*
97. Sopwith 7F.1 Snipe*
98. Sopwith F.1 Camel*
99. Sopwith Pup*
100. Sopwith Triplane*
101. SPAD XIII*
103. Boeing 100/P-12
106. Messerschmitt Bf 109E-3
107. Stearman PT-13A Kaydet
108. Aeronca L-3 Grasshopper
109. Beech C-45H Expeditor
110. Boeing B-17F Flying Fortress
111. Boeing B-29 Superfortress
112. Curtiss P-40N Warhawk
113. General Motors FM-2 Wildcat
114. Goodyear FG-1D Corsair
115. Goodyear F2G-1 Corsair
117. Lockheed F-80C Shooting Star
118. Lockheed P-38L Lightning
120. Nakajima Ki-43-IIb Hayabusa
121. North American P-51D Mustang
122. Pratt-Read PR-G1
123. Republic P-47D Thunderbolt
124. Supermarine Spitfire Mk.IX
125. Yakovlev Yak-9P
126. Boeing B-52G Stratofortress
127. Boeing VC-137B
128. Boeing WB-47E Stratojet
129. Canadair CL-13B Sabre
130. Fiat G.91 PAN
131. Globe KD6G-2 Firefly
132. Grumman F9F-8 Cougar
133. Mikoyan-Gurevich MiG-15

134. Mikoyan-Gurevich MiG-17F
135. Northrop YF-5A
136. Vought XF8U-1 Crusader
138. Douglas A-4F Skyhawk II
139. Lockheed D-21B
140. Lockheed F-104C Starfighter
141. Lockheed M-21
142. McDonnell F-4C Phantom II
143. Mikoyan-Gurevich MiG-21PF
144. Mikoyan-Gurevich MiG-21PFM
145. Lockheed YO-3A Quiet Star
146. Boeing AGM-86B ALCM
147. Cessna 0-2A
148. Grumman A-6E Intruder
149. Grumman F-14A Tomcat
150. Lockheed MQM-105 Aquila
151. McDonnell Douglas F/A-18C
152. British Aerospace/McDonnell Douglas AV-8C Harrier II
153. Lockheed Martin/Boeing RQ-3A DarkStar
154. Boeing-Insitu ScanEagle
155. Sputnik
156. McDonnell Mercury Capsule**
157. Rockwell Apollo Command Module
158. Boeing Inertial Upper Stage
159. Boeing Lunar Roving Vehicle**
160. Resurs 500
163. Piasecki H-21B Workhorse
164. Sikorsky HH-52 Seaguard
165. Bell UH-1H Iroquois
167. Viking
169. Soyuz-TMA 14
170. NASA Full Fuselage Trainer (FFT)
171. Blue Origin Charon

*reproduction or replica
**mockup

Lockheed M-21 Blackbird in the Great Gallery.

Zero to Fifty

The Museum of Flight is a vibrant part of its community and a national treasure. During its first 50 years of ups and downs, the organization has forged a leadership role. What was learned, what seems to work, and what does all this mean for the future of flight museums?

Vision

Vision is not a static exercise, something anchored in place; it shifts and changes with new possibilities, circumstances, and participants. It is the discipline of setting high standards and looking ahead to anticipate alternatives, to avoid problems, and to find better solutions. It is a process that responds to unexpected opportunities and surmounts challenges.

The MOF has had visionaries. They thrive in the organization, finding new approaches to problems blending far-out possibilities with near-term realities. This is the vision thing—the persistence to keep a good idea alive until finding a practical means of implementation.

The organization has flexed to stay on the visionary course, locating properties, adding to the collection, and building structures, with a scheme of growth unthinkable in the landlocked early days. With success, the vision thing is a constant component of the team structure.

Mission

No task is more important to museum builders than preparing a solid statement of mission. This commitment is packed with significance, a few words that define the institution to meet the community's need. Based upon this declaration are the museum's size, collection, location, and all other guiding parameters.

No matter how well crafted, the mission must be regularly evaluated. The MOF has learned to correct course and realign objectives. Over several iterations throughout the past 50 years, the mission has been clarified, yet the emphasis is always on education. With interplay between the mission and a changing environment, the museum has been true to course in its navigation.

Organization

In manufacturing as well as in sports, teamwork is an essential organizational model. A coordinated team generates better products and wins games, although setting up such an organization is not simple. The traditional museum nonprofit model is to elect a governing board that sets policy, adopts budgets, and oversees financial issues while retaining a CEO to manage staff and operations.

How can a large organization operate as an efficient team? The MOF has developed a strong committee structure in support of governance whose members have distinct experience.

Board chairs have previously worked committee assignments, some serving on virtually all. This familiarity, coupled with a passionate interest in the purpose of the institution, builds capable leadership. Each year attracts new talent, as well as more knowledgeable trustees who can interact effectively with the CEO and staff.

A nonprofit board is a fiscally diligent body that accepts a public duty for service and a fiduciary responsibility to donors. Everyone who serves as a trustee takes this seriously. Along with this, there is also a strain of risk-taking, the willingness to act quickly to seize opportunities, to take on debt when necessary, and to commit to a priority project even before the resources are in hand.

It is not uncommon to have an MOF board member personally make a significant financial contribution, then assist in implementing a task. This kind of overlap in roles is not interference. Committees are stronger when their members are personally involved in implementation. The team as a "committee driven" structure is a formidable strength that adds to the viability and sustainability of the institution.

Location

Building a flight museum at Boeing Field was controversial. The fact that airport management, tenants, and others initially opposed its location is now inconsequential. This airfield idea challenged the status quo, a bias of museums and airports at that time. It took persistence to demonstrate what seems so obvious today.

The seven-acre museum property on Boeing Field was valued conservatively at $3 million at the time of construction. A modest sum as one looks back, but monumental. King County was asked to make the site available at no cost, a winning proposal ultimately negotiated over several years of studies and hearings. This was not just a bold request for public support, but a breakthrough with national implications for museums to become a recognized airport-related use. The active airfield site has proven a crucial decision for the eminence of the Museum of Flight.

Concept

The concept for the MOF complex of buildings was years in the making. The museum's first program of requirements was to provide a full-service educational institution covering all aspects of aviation and space. This commitment was made well before owning enough artifacts to pull it off. The physical facilities were planned and built to take advantage of the airfield location and meet the institution's needs.

As a cultural institution in an industrial neighborhood, the organization was to plow new ground. Having been warned that no one would visit Boeing Field for anything but their jobs and to fly, the team was tasked with creating an oasis in the industrial environment. The museum reached out to the business, tourism, and hospitality sectors, who responded with essential advice and support. The result was a cultural and educational destination of tremendous value to the city, the state, and the nation.

Artifacts

Everyone has a favorite artifact. Reasons may vary, but surveys indicate that almost every item in the collection has an advocate and a story. These are the celebrities, some more attractive than others, some leading, some supporting, but each playing a role in the story of the Museum of Flight.

Former trustee and aviation great Ernie Gann was adamant in suggesting that the museum should have some stars that set the stage for a dramatic succession of acquisitions. Each artifact, not just the rarest or the largest, shaped the institution. Some transcend luminescence and exert influence on museum development.

The 80A, along with the Boeing 247D, gave early proponents historic assets to make their case. The magnificently restored Curtiss JN-4D commanded respect for an institution in its infancy. The journey to find a B-17 is an important story of human interest. This rare F model, built in Seattle, and its generous donor exemplify the spirit of the museum. The 747 jumbo jet, a priceless first of its kind, initiated the need for a large facility for display and interpretation. The triumphs in acquiring other rare and priceless artifacts, military and commercial, many first editions, accelerated this need.

The collection of space artifacts has grown at a pace not considered possible at the onset. Now, with the addition of evolving private space initiatives, the Museum of Flight offers a compelling story in its new Space Gallery.

Working for a collecting museum, what could be more important than artifact preservation? The MOF scores high marks on this obligation with its restoration center, protective environment, and high quality of interpretation.

Archives

The Great Gallery incorporated a library and archive space proportionately larger than that of most flight museums but soon reached its capacity, and a larger, dedicated structure was created on the west side of the campus. Over time one success begat another, and the cycle of an exceptional facility, great artifact collection, and growing archive became proof of a kind of cultural ecosystem.

The MOF has proved competitive and insightful in acquiring rare and precious documents. Supporters see the value of seeking these important materials and then purchasing them or cultivating donations. This collection of historical documents has grown to be one of the largest in the world. The aerospace archives are the true soul of the museum.

Fundraising

From the initial "friends and family" period of financing the startup, the museum has engaged in campaigns for construction funding. Once in operation, the institution practiced all of the fundraising categories, from annual giving to operations to periodic capital campaigns, targeted gifts and grants from private and public sources, and planned giving.

The museum has evolved into a continual capital-acquisition process with both annual and long-term objectives. Its galas help to enrich this process. This system integrates with museum development initiatives that are reaffirmed each year.

For the Museum of Flight, the equivalent of the Great War was the campaign to build the Great Gallery and complete the original campus. Dubbed *For Future Generations*, this campaign was slow to organize, challenged to attract leadership, and in turn limited in its drafting of foot soldiers. It took some time to gain momentum.

It had starts and stalls, more than are typical. It also set the largest target for a capital investment of any museum in the region up to that time. Not only was it a demanding objective, but its implementation always seemed to occur during a weak economy. The effort went national in scale yet also brought in many small gifts, both factors contrary to most local fundraising, and not a sustainable strategy.

The lesson is that campaigns are not occasional and discrete; they are continuous and integrated. Major maintenance, purchases, collection acquisition, land, buildings, exhibits, and yes, support to annual operations will require capital. This harsh reality must be embraced in the organization as it matures. With this approach, the museum team can manage to fundraise annually and avoid unrealistic deadlines. This model, in effect at the MOF, is integrated with an organizational structure that features committee involvement and action.

A corollary feature of this fundraising process is the integration of annual galas to generate funds for education, celebrate historic events, and build awareness. Long past the days of bake sales and dances, the museum galas have become must-attend cultural activities in the community.

In the mid-1990s, board member Jim Curtis suggested that the museum was a prime venue for a cultural auction. Seattle had distinguished itself with this form of fundraising, particularly with the legendary PONCHO auction.

Dick and Sharon Friel became famous auctioneers at PONCHO and elsewhere in the region and beyond, taking this entertainment/fundraising theme as far as Australia. Dick Friel was an aviation enthusiast and marketing manager who had a foot deeply set in show business. As an auctioneer, he ensured that guests would have fun, which produced dollars. Dick and Sharon knew aerospace, did not hesitate to engage in this new venture, and stayed involved in all of the museum's auction parties.

From the first, these galas proved eminently successful. Sponsors and donors immediately rallied to the cause. Though growing so large as to require a performance tent on the museum grounds, these galas continue to have delightful moments under the aircraft in the Great Gallery. At the first of these auctions in 1997, astronaut Bill Anders donated a flag that had encircled the Moon on Apollo 8, which attracted a bid of $125,000, at the time the highest bid offered for any item in Seattle. That bid was surpassed in subsequent auctions.

The galas and auctions generate net revenue to support operations and specific education programs. They also attract supporters and donors from beyond the aerospace community, enhancing sustainability.

Operations

With a new museum, what happens next? You have the keys, and it is near perfect. So what happens day in and day out?

The transition from building to opening doors for a surge of crowds is challenging. The museum finds that cash-flow issues and capital requirements emerge immediately. The euphoria of grand opening crashes into the reality of demands for sustainability.

The MOF has rallied to this challenge effectively through each of its expansions. In early years, the museum steadily built its programs and public service in scope and quality while meeting the annual budget from earned income. A large membership, profitable gift shop, and special events augmented the admission fees.

The museum's financial condition was shaky at the outset but improved remarkably. There was steady growth in net worth, from $14 million in 1984 to $30 million in 1989, and then it more than doubled over the next decade to $85 million, increasing rapidly to $233 million in 2014. Once stabilized, the museum grew in asset value and its ability to attract capital. The institution also cultivated additional leadership with the capability to acquire resources, and this has continued through ups and downs in the regional economy. Financial security and accountability were then and still are high priorities of the Foundation. Budgets based on annual operating plans and potential revenue projections are carefully reviewed for performance. The museum focuses on sustainability over the long term.

The museum has consistently followed professional business procedures. Problems are identified and addressed. The experience informs the process. What works is improved and what does not is fixed. At the MOF, good business is a discipline taken seriously, and it has helped the institution to survive and thrive during several economic downturns.

Visitor Services

A museum is ultimately judged by how well it serves its visitors in convenience, safety, and comfort. Only then can it be as memorable and educational as it should. All of the work incorporated into the institution is lost if the core amenities are missing.

From its opening, the MOF organized the operation with a visitor-service orientation. The reward is better public service as well as healthy economic performance. Today in the museum world, this visitor-service orientation is *de rigueur*.

Volunteers

The definition of a museum team begins with its volunteers. The original founders were all volunteers, as are its trustees; they neither expect nor are given financial compensation. The tasks they perform are for the benefit of the institution. The trustees are the first volunteers, but not the only ones who give of their time and resources. Volunteers are evident in all levels of museum operations.

One of the treasured resources of the MOF is this base, which has flourished over the years. The willingness to give back is ingrained in the aerospace community, as well as those who represent other skills and professions. This resource is essential.

Among the most noticeable are the docents, the knowledgeable interpreters who work the gallery floor greeting visitors, leading tours, and answering questions. Bringing experience and expertise to the museum gallery, they commit to a substantial time obligation and are trained to understand the mission and core values. The MOF is fortunate to have a force of hundreds of docents, and their grasp of flight subjects is deep and vast.

The docents are front and center, but volunteers support every part of the museum, every function, every division. Artifact restoration is flush with volunteers, often providing skills and experience that would be difficult for the museum to find otherwise or afford. The story of MOF development is punctuated with transitional points when a volunteer effort saved a valuable artifact and advanced the mission. For some of the large galas and special events, hundreds step up to serve at one time.

It would be very hard to estimate what value these efforts add to the museum, but it certainly would be in the tens of millions of dollars. One can go on and on with stories of volunteer service from earliest days to the present, but it's hard to find the appropriate level of praise.

Why do they fulfill their tasks with enthusiasm, receiving no financial compensation and little glamour? Their hours are long, at times a slack period interrupted by frantic, crowded sessions with large groups of sometimes impatient visitors. Those who serve say it is a privilege to volunteer.

If Alexis de Tocqueville were to return to Earth for a look around, he would instantly recognize the Museum of Flight as typically American, full of idealism, with people solving problems, folks filling needs, the spirit of volunteerism everywhere. He would recognize the great value of this institution, not just in a multimillion-dollar physical facility, but also in the spirit of community service.

Connections

Puget Sound is an aviation center, and there are generations of connections with flight history. Thousands of people have personal or family stories that relate to the mission of the museum. Some personal connections reach out and expand over the years, adding to museum sustainability. In the contemporary network society, personal and professional relationships define us. So it is at the Museum of Flight.

The aerospace community is a constellation of global influence, and the flight test community stepped up early in planning. For whatever the reason, these people were ready to preserve the history and technology and to build a museum.

Elliott Merrill, the first chair of PNAHF, was a heralded manager of test flight for The Boeing Company at the project's infancy. Harl Brackin enlisted him to the cause, but the recruitment was an easy assignment because Merrill was a believer. He knew everyone at Boeing, but also many enthusiasts outside the company.

When Dick Taylor returned from WWII duty and signed on at Boeing as an engineer, he had his sights set on flight test. Elliott Merrill hired him into this division, where Taylor met the seasoned test pilot Clayton "Scotty" Scott, who would become a mentor to Taylor. Scotty, in his work with attorney Bob Mucklestone, encouraged him to take up flying, which he did with a passion. When Mucklestone asked Scotty to join PNAHF, Scotty referred him to his friend Dick Taylor. Mucklestone met Taylor and recruited him.

Dick Taylor recalled that as The Boeing Company's 50th anniversary approached, the corporate planners decided to contract for a replicated B&W as a centerpiece. Taylor suggested the replica would best be done in the Seattle area and recommended his good friend Scotty, retiring from flight test to operate an aircraft modification business at the Renton Airfield, as the right contractor for the job. Scotty and his firm, Jobmaster, were awarded the contract and built the replica in Renton, involving a lot of Boeing old-timers as volunteers. The B&W and its resulting flights were highlights of the 50th anniversary, which PNAHF also supported. The replica B&W was first loaned and then gifted to PNAHF and is now an important feature of the MOF.

When Dick Taylor was posted to Washington, D.C., to run the Boeing office, he called Bob Mucklestone to offer his resignation from the PNAHF. Mucklestone countered that a resignation was not necessary and suggested that Dick could be helpful in his new position. The connection from Merrill to Scott to Mucklestone to Taylor served the project well. It was serendipitous and unplanned, but an important web of personal relationships.

Another interrelated set of acquaintances occurred with the volunteer efforts of Jack Pierce. Pierce lived on a houseboat and met a neighbor, David Williams, who was a design architect with Ibsen Nelsen and Associates and married to Nelsen's daughter. In the process of collecting his volunteers, Pierce tagged Williams to get Nelsen's office involved. Both Nelsen and Pierce had connections to Seattle's cultural scene, and Nelsen had numerous contacts in the design community and with elected and appointed officials. This nucleus quickly formed an assemblage of promoters and supporters that enhanced the proposal for a flight museum in Seattle.

Among Nelsen's contacts was young B. Gerald Johnson, who at the time had risen to the position of administrative assistant to U.S. Senator Warren Magnuson. Johnson had gone to school with one of Nelsen's children and knew and respected the architect for his involvement in historic preservation and urban planning.

With his position in the Capitol, Johnson worked with Jack Pierce on Boeing Company matters. He was also able to make additional connections for PNAHF planners. When the Red Barn was unexpectedly scheduled for deregistration as a historic place, purportedly because of its move to Boeing Field, PNAHF scrambled to oppose the action. Gerry Johnson set up a meeting in Washington, D.C., with top officials of the National Park Service. The meeting resulted in an immediate retraction of the proposal.

On a later occasion, Johnson used the offices of Senator Magnuson to set up meetings with the State's congressional delegation. Those conversations resulted in Senator Magnuson and the delegation supporting the project and offering a federal funding approach with requisite requirements for local private and public match. Although this venture did not materialize, it was a good sign for the credibility and maturing of the project. Upon his return to private legal practice in Seattle, Gerry Johnson continued to advise the project, taking the lead in some successful strategies for leasing land, financing construction, and establishing official liaison with county government.

Another set of connections is that of Bruce McCaw, Joe Clark, and Clay Lacy and their friends. McCaw and Clark were fellow aviation enthusiasts. Both crossed paths with Lacy during one venture or another, and all came to be close friends. Each had a passion for flight, but also a discrete and complementary set of interests and experiences. In particular, their friendship and ventures with Bill and Moya Lear were to prove advantageous to the museum. Bruce McCaw's father served on the Lear Jet board, forging a lifetime friendship between the families. Joe Clark also had many business relations with the Lears and Lear Jet. Lacy also worked with the Lears, demonstrating the Lear Jet to influential people and introducing the aircraft to a world market. Lacy has had a legendary flying career that spans decades and logged as many hours flying as anyone on Earth. He has set at least 29 world speed records in various aircraft. Lacy was a pilot in 1962 on the first flight of the Boeing 377 Stratocruiser "Pregnant Guppy," the aircraft modified to carry the Saturn rocket booster. Clark and Lacy have flown several record-setting flights demonstrating Clark's Aviation Partners winglet technology. The experiences of these friends often intersect. Their comradeship and set of connections have resulted in many contributions to the museum. Their influence continues.

Bruce and Joe took an early interest in motorsports, cars, hydroplanes, aircraft, almost anything that was fast, which tied them into a mutual friendship with Charles A. "Chuck" Lyford III, another local boy who grew up racing. Lyford was a boyhood friend of McCaw and an ardent and champion hydroplane and aircraft racing competitor. Lyford was the first of the three to serve on what was then the PNAHF board, and his son Charlie would later serve as well, chairing the Collections Committee. These longtime friends can tell story after story of joint projects or occasions when their interests would bring them together. Their separate connections extend this universe outward. These personal relationships and knowledge of the aerospace community are a considerable asset to the museum.

Some associations sprang up during service to the institution. A particularly advantageous example is that of Gene McBrayer and Ed Renouard, who met in service on the board and came to blend their complementary talents, garnering the title "the tag team" for their organization of museum building and construction projects. This relationship has provided extraordinary leadership in support of the largest of museum projects, one-off structures, keeping them on schedule and often helping to bring them in under budget. Their connection has resulted in an invaluable service to the museum.

And then there are the family connections. Former trustee D. P. "Van" Van Blaricom was attracted to the project by his personal ties to flight, the Red Barn, and Boeing Field. His grandmother who raised him worked in the Red Barn and was one of the women in the classic photo of seamstresses sewing fabric on wings in the factory. Her younger sister's husband was in the family that owned the Prentice Nursery on the land where the museum is now located.

Philip G. Johnson, son of an early Boeing Company president, was a museum trustee from 1976 until his death in 2013. He was married to Marcia Johnson Witter, hired as PNAHF's first education consultant in 1977. Their son Matt trained as a docent with his father at the age of 10, becoming the youngest to volunteer at the Museum of Flight, and his brother, Alex, was a counselor at Space Camp.

In keeping with family tradition, several fathers and sons have served on the board of the museum, including founding trustee Harold "Kit" Carson and his son Scott, Dick and Steve Taylor, John Fluke Sr. and Jr., Gerald and Keith Grinstein, and David and Keith Wyman. Father and daughter teams include Eddie and Jane Carlson (Williams), and Hunter and Anne Simpson. Husbands and wives who have served on the board include Ernie and Dodie Gann, and Ned and Kayla Skinner. Brothers James and John Nordstrom are also board veterans. These family connections create equity that is expected to survive over generations.

Finally, over the entire 50 years, there has been an essential connection between William E. Boeing Jr. and The Boeing Company. Initially, the company was a family-controlled corporation. Later, with the Boeing family separated from the business, William E. Boeing Jr. and the company cooperated in saving the historic Red Barn.

In more recent years, both were honored with the first of the Red Barn Heritage Awards, bringing them closer. Finally, the recent joint announcement of the $30 million contribution to establish the Boeing Academy for STEM Learning brought each back to an essential partnership. This Boeing connection from William E. Boeing Jr. to the corporation to joint investment is an interesting link that has been of critical value to the institution.

Action

The Museum of Flight is an exceptional organization, not due to mere good fortune or a blank check. Somewhere between vision and reality, there was luck, but also plenty of hard work, conventional wisdom, and unconventional strategies.

The museum's blue-sky visionaries proved to be pragmatists in disguise who kept ideas alive long enough to make them practical. The leaders kept pushing as long as possible, then, before fading, passed the baton to the next. Dreaming became a component of implementation.

The pioneers of PNAHF, the proud, the few, sometimes referred to as the "strut pluckers," did welcome the downtown "bankers," yet did not entirely trust folks who did not fly. But this was all for a worthy cause, and to save some damn fine old aircraft, and to educate kids.

Sometimes the most obvious needs are lost in plain sight. Of course Seattle should have an aerospace museum. It all began to make some sense. It took a decade to get organized, yet in a novel way that mined aerospace connections, in which donors implemented, volunteers worked as staff, and employees contributed time and resources. Whether top down, bottom up, or inside out, the MOF has found a way that works.

Who would even consider locating a cultural attraction at a noisy airport smelling of aviation gasoline? Who would entertain the idea of replacing the Circle Tavern and an abandoned trailer court with a museum in an industrial area? Who would ask King County to be a co-conspirator and then ask them for the money to do the deed?

As if that gambit was not enough, how about having the nerve to suggest a big steel-and-glass gallery in the midst of public works and factories? The MOF wanted a new kind of display gallery —one that an aviation trade magazine called "a tribute to bird droppings"—although widespread opinion was that an ideal location for a flight museum was in an old, dark hangar. The MOF built that elegant structure. What at the time seemed so far out and preposterous is now taken for granted. That is what can happen with practical visionaries, risk-taking governance, and a dose of theater.

The MOF has a lot of moving parts. The museum has survived fundraising and board building and endless meetings to enjoy construction, collecting, and operating to a mighty mission. The campus vibrates with programs and activities for lifelong education. Priceless artifacts of air and space history abound, with something for everyone. The archives amaze researchers around the globe. The institution has scratched up $200 million in cash contributions, with at least that much more in volunteer and contributed services. Every day the assets grow, and the return on investment is enormous by any measure. The result is exceptional service to millions of visitors and participants. Each year hundreds of thousands of students are inspired, educated, and prepared for careers and life. The museum works.

All planning must lead to taking action. The ability to act quickly is not inherent in nonprofit governance, but such agility can make a difference between good and great, winning and losing, and ultimately in fulfilling the mission.

Alan Shepard in his Freedom 7 Mercury spacecraft suffered so many hours of delays that he urinated in his space suit. Finally, agitated, he shared with mission control: "I'm cooler than you are. Why don't you fix your little problem and light this candle?" He went on to suborbital flight as the first American in space.

So it is at the Museum of Flight. The mission is demanding. The team is experienced, trained, and well prepared. It is big, but agile, well coordinated, and efficient. When the time comes, it cuts through delays and sends the message to "Light This Candle."

Acknowledgments

Apex Foundation and Bruce R. McCaw have been generous with their time and financial support and leadership to make sure that this story was told. Bruce is a longtime volunteer, strategist, donor, and committed keeper of the flame. His personal records of service to the museum are copious and benefited this storytelling. He maintained a commitment to the history, gathering additional information throughout this process of research and writing. I think no other individual knows more about this museum than Bruce. He was there. The museum is better for his generosity of spirit and resources.

Craig Stewart of Apex Foundation has been a constant supporter and therapist and friend. He, too, has a long service with the museum, and he cares about its history. Aurelie McKinstry of Apex added her calming influence to the team.

My wife, Peggy Nuetzel, should have her name on the cover of the book. She assisted me with writing, editing, and most of the research. Peggy's volunteer service with the museum began in 1980, and her recollections and photographs were invaluable in recording the institution's back-story. Peggy's education in journalism brought needed discipline to this task. I could not have done this without her.

Our friend Jay Spenser added a lot with his definitive sidebar materials. He was a curator at both the National Air and Space Museum and the Museum of Flight. Now retired from Boeing and busy with his career as an author of fiction and nonfiction, Jay is a writer who truly grasps aviation history.

Volunteers, board members, staff, and donors have given freely of their time to help record this history. On and off the record, all spoke with candor and expressed enthusiasm for the museum and its meaning to the community. It almost became a mantra to hear one of these players state emphatically that this service was a privilege and that the museum meant more to them than they to it. This institution of 50 years has made a lot of friends along the way.

Everyone did what was possible to help the team with this story. Museum staff assisted within busy schedules. The publisher, Documentary Media, spared nothing to support the effort. They care about recording organizational histories in our community, and as is so often the case, we found they had an abiding interest in aerospace. Editorial Director Petyr Beck and his team of Editor Judy Gouldthorpe and Designer Paul Langland were patient with the writing and supportive of our needs.

Alison Bailey was the museum staff coordinator, and no one could be better at that job. She has served the Foundation and its members for more years than seem possible, given her youthful looks. Alison is human institutional history, and we all are lucky to have her to help keep the record straight. Rosie Gran somehow found time within her museum communications balancing act to help us—to reserve a meeting room, or check a schedule, or find coffee. The staff of the museum archives and Ted Huetter in public relations assisted with photos and data.

Peggy and I are humbled by this overwhelming task. Inevitably, stories will be missed, important events will be lost in the recording, and the errors and omissions will loom large upon reading. The best one can hope for is that the story elicits a new conversation about this history. The Museum of Flight is 50 years old, but it is just getting started.

— Howard Lovering

Board Chair

A. Elliott Merrill*

Chairman   1965-1973
Trustee   1965-1984

Profession and Affiliations: Boeing test pilot; first Chairman of the Washington State Aeronautics Commission; founding trustee of the Pacific Northwest Aviation Historical Foundation (PNAHF); received the Pathfinder Award in 1985 in the Flying category

Highlights during Chairmanship:
The Museum of Flight's founding Chairman of the Board
The museum receives its 501(c)(3) status with the State of Washington
A 1929 Boeing 80A is airlifted from Alaska to Seattle on a C-124
A 1933 Boeing 247D is acquired in Bakersfield, California, and flown to Seattle by Jack Leffler and Ray Pepka
The City of Seattle leases a building near the Space Needle at Seattle Center to PNAHF for an aerospace exhibit
The Boeing B-47 Stratojet is acquired from the military and flown to Boeing Field

Thomas R. Croson*

Chairman   1973-1974
Trustee   1965-1981

Profession and Affiliations: Vice President of Community Relations at West Coast Airlines; airline history and operations consultant; pilot; founding trustee of Pacific Northwest Aviation Historical Foundation

Highlights during Chairmanship:
The Boeing Red Barn is listed on the National Register of Historic Places

James Dilonardo*

Chairman   1974
Trustee   1969-1984

Profession and Affiliations: OX5 Club; historian; Display Supervisor at Sears Roebuck & Co.

Highlights during Chairmanship:
Ibsen Nelsen and Associates, an architectural firm, is introduced to PNAHF through Boeing's Jack Pierce
Museum planning and conceptualizing begins

Robert S. Mucklestone

Chairman   1974-1977
Trustee   1969-2013

Profession and Affiliations: Partner, Perkins Coie law firm; pilot; obtained and restored a Boeing P-12 for the museum with Lew Wallick and Orville Tosch

Highlights during Chairmanship:

King County signs agreement to relocate the Red Barn. It is barged up the Duwamish River and trucked to its current location at Boeing Field
Open house at the Red Barn draws 20,000 people
Education program is initiated
Howard Lovering, an urban planner from Boeing, is hired as the first full-time Museum Director

George C. Briggs*

Chairman   1977-1979
Trustee   1975-1998

Profession and Affiliations: Athletic director, University of California and University of Washington; Banking executive at SeaFirst and First Interstate Bank

Highlights during Chairmanship:

The museum concept is revised to house 20 major aircraft in a proposed "Great Gallery"
Museum membership grows to 2,000
Aviation-based education programs are taken to 66 schools statewide, and an impressive 3,000 students participate

Richard E. Bangert*

Chairman   1979-1988
Trustee   1977-2004

Profession and Affiliations: CEO and Chairman, First Interstate Bank of Washington; civic leader; private pilot
Longest-serving museum chair

Highlights during Chairmanship:

The museum's Board of Trustees adopts a building program to renovate the Red Barn and build the Great Gallery
Boeing Chairman T. A. Wilson presents a $1 million gift from The Boeing Company for renovations on the Red Barn in honor of William M. Allen
Boeing Company employees contribute nearly $1 million to the museum's Red Barn renovation project
The newly refurbished Red Barn opens to the public in September 1983
T. A. Wilson, Chairman of the Board & CEO of Boeing, accepts chairmanship of a new $26 million capital campaign to build the Great Gallery
127,000 people visit the museum's Red Barn in its first full year of operation
Officials break ground for the Great Gallery in a highly publicized ceremony
The Museum of Flight receives official "accreditation" from the American Association of Museums
The museum's Great Gallery officially opens to the public on Sunday, July 12, 1987

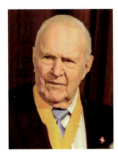

Mark E. Kirchner*

Chairman   1997-1999

Trustee   1986-1999

Profession and Affiliations: Boeing engineer and executive; pilot; aircraft builder

Highlights during Chairmanship:

The museum holds its first major fundraising auction, which raises over $1 million for educational programs

Dr. Henry Kissinger dedicates the first presidential jet, Air Force One, and it officially opens to the public

The Capt. Elrey B. Jeppesen collection is acquired

A new permanent exhibit, *The Tower*, opens and is dedicated to the memory of Mrs. Bertha Boeing and
    Thorp Hiscock

Harold E. Carr

Chairman   1999-2001

Trustee   1990-present

Profession and Affiliations: Vice President, Public Relations & Advertising, The Boeing Company; member of numerous nonprofit boards

Highlights during Chairmanship:

The Douglas DC-2 is acquired

The Education Department, buoyed by the museum's successful galas, grows to 14 full-time and 20 part-time educators
    and offers 18 different core programs to kids in grades K-12

The museum holds its second major auction, the "Out of This World for Kids" gala auction, in July, raising $2 million
    for children's programs

A new major exhibit, *Rendezvous in Space: A Tribute to Pete Conrad*, opens in 2000

David C. Wyman

Chairman   2001-2002

Trustee   1992-2013

Profession and Affiliations: Private investor, Wyvest; Chairman, Rossoe Energy; Partner, Wyman Lumber Company; Chairman, Item House; community leader; pilot; member of several corporate and nonprofit boards

Highlights during Chairmanship:

The "Sky Without Limits" capital campaign is initiated to raise funds for a new wing to house the
    Champlin Fighter Plane Collection

Personal Courage Wing is conceptualized

Moya Olsen Lear – wife of the late Bill Lear – donates The Lear Archives to the museum

H. Eugene McBrayer

Chairman   2002-2003

Trustee   1997-present

Profession and Affiliations: President, Exxon Chemical Company, and industry executive; pilot; aircraft builder

Highlights during Chairmanship:

Children's participation in museum programs grows to 80,000 for the year

The museum breaks ground on the new Personal Courage Wing

The library and archives are moved into the former Boeing 9-04 building – a facility that was donated by
    The Boeing Company

A new exhibit, "Building for Air Travel," opens in the Bill and Moya Lear Gallery

Bruce R. McCaw

Chairman   2003-2005

Trustee   1985-present

Profession and Affiliations: Chairman, Seattle Hotel Group; Co-Chair, Apex Foundation; Chairman Emeritus, Pistol Creek Co.; Board of Directors, National Air and Space Museum; Co-founder, Horizon Air; Former Chairman, Westar Insurance; VP and Director, McCaw Communications Companies; Co-Chair, Talaris Foundation; Owner, PacWest Racing Group; Co-founder, Friendship Foundation; pilot; philanthropist; aviation historian; received the Pathfinder Award in 2013 in the At-Large category

Highlights during Chairmanship:

The world's first Boeing 737 makes its final flight to join the museum's collection

The museum forges a partnership with Highline School District's Aviation High School

A priceless collection of never-before-seen Wright brothers documents is acquired

The Champlin Collection of World War I and World War II fighter aircraft is moved from Mesa, Arizona, to Seattle

The museum acquires British Airways Concorde G-BOAG. It is flown into Boeing Field in 2003 with incredible
    media and public attention

A newly renovated lobby welcomes visitors and is named in honor of the late Keith W. McCaw

The museum's Aviation Learning Center – an innovative educational experience – is opened and sets a new standard
    for museum-based interactive learning

The J. Elroy McCaw Personal Courage Wing opens to the public. As part of the opening, a thousand people attend
    the "Wings of Heroes" gala and view the galleries for the first time. The gala honors members of the American
    Fighter Aces, the Flying Tigers, Tuskegee Airmen, and Medal of Honor recipients

Membership climbs to 22,000 annually

James T. Johnson

Chairman   2005-2007
Trustee   1999-2011

Profession and Affiliations: VP & General Manager, Everett, The Boeing Company; Executive at Pratt & Whitney; President, GE Capital Aviation Services; President & CEO, Gulfstream Aerospace Corp.

Highlights during Chairmanship:

The museum refocuses its educational goals to align with national science, technology, engineering, and math (STEM) standards

Exhibits in the Red Barn are updated, enhancing the visitor experience in Boeing's first factory

The museum announces the launch of a new educational program, Washington Aerospace Scholars, to expand the museum's efforts in STEM education

The Kenneth H. Dahlberg Military Aviation Research Center opens

The museum hosts its first "blockbuster" exhibit with *Leonardo da Vinci: Man, Inventor, Genius*. The traveling exhibition generates big revenues and large crowds

Robert J. Genise

Chairman   2007-2009
Trustee   1998-present

Profession and Affiliations: CEO, Boullioun Aviation; CEO, Dubai Aerospace Enterprise; CEO, Aergen

Highlights during Chairmanship:

Bill and June Boeing donate a historically accurate Boeing Model 40B airmail plane that goes on display in the Great Gallery

A new exhibit, *Space: Exploring the New Frontier*, opens to the public

The museum acquires 6.47 acres on the west side of East Marginal Way S. for future expansion

Plans get under way to connect the museum's east and west campuses with a pedestrian skybridge

Museum educators reach more than 120,000 youths both on-site and through outreach programs

The T. Evans Wyckoff Memorial Bridge is completed and provides an unparalleled level of safety for museum visitors, connecting the museum's east and west campuses

J. Kevin Callaghan

Chairman   2009-2011
Trustee   1997-present

Profession and Affiliations: Chairman, Badgley Phelps Investment Managers; pilot

Highlights during Chairmanship:

A 1954 Lockheed Super G Constellation arrives after a long journey by land, and is added to the museum's permanent aircraft collection

The museum breaks ground on a new space gallery to house NASA's Full Fuselage Trainer

K-12 education reaches an all-time high of 140,000 participants

A meticulously restored P-51D Mustang is installed in the Personal Courage Wing

The museum receives a highly coveted reaccreditation from the American Association of Museums

Michael R. Hallman

Chairman   2011-2014
Interim CEO   2010
Trustee   1990-present
Profession and Affiliations: Sales & Marketing executive, IBM; Chief Information Officer, The Boeing Company; President & COO, Microsoft

Highlights during Chairmanship:
The museum launches a two-year strategic planning process entitled *Vision 2020*
Construction is completed on the Charles Simonyi Space Gallery
The inaugural Red Barn Heritage Award is presented to William E. Boeing Jr.
The museum sets an all-time attendance record with 540,346 visitors
Over 156,000 youths participate in a museum-directed educational program – also setting a new record
Charles and Lisa Simonyi donate the Soyuz TMA-14 capsule for display in the new space gallery
The museum acquires a UH-1H Huey helicopter to enhance the Vietnam exhibits
Blue Origin loans its first vertical takeoff and landing test vehicle, Charon, and it goes on display in the
     Simonyi Gallery
A three-year, comprehensive fundraising campaign entitled *Inspiration Begins Here!* is launched to realize the initial
     goals of *Vision 2020*. Goal: $77 million
The William M. Allen Theater is renovated, including new state-of-the-art projection, sound, and seating
Museum trustees and "friends of Amelia" embark on a fundraising drive to acquire a Lockheed Electra 10-E.
     The beautiful Electra arrives and is put on permanent display

William S. Ayer

Chairman   2014-2016
Trustee   2002-present
Profession and Affiliations: Chairman & CEO, Alaska Airlines and Alaska Air Group; Chairman of the *Inspiration Begins Here!* comprehensive campaign; received the Pathfinder Award in 2012 in the Operations category

Highlights during Chairmanship:
Inspiration Begins Here! Campaign surpasses the 87% mark with $67.3 million raised
The Boeing Company presents the museum with a 787-8 Dreamliner, which immediately goes on display
     in the Airpark
The Michael and Mary Kay Hallman Spaceflight Academy opens to wide community and industry acclaim
The Boeing Company receives the second Red Barn Heritage Award, an event chaired by Bill and June Boeing
The museum embarks on a comprehensive educational initiative to grow on-site outreach and accredited programs
     and increase student access to STEM career pathways – particularly for underrepresented youths
The Alaska Airlines Aerospace Education Center opens
The Boeing Company and June Boeing join together to announce their unprecedented matching gifts of $15 million
     each to further the museum and its role in the future of STEM education, and the Boeing Academy for
     STEM Learning is born
Jeff Bezos unveils components of the Apollo 12 and Apollo 16 Rocketdyne F-1 engines, a gift made possible by
     NASA's Goddard Space Flight Center and Bezos Expeditions
Construction begins on the immense 3.2-acre Aviation Pavilion on the museum's west campus

Anne F. Simpson

Chair   2016-present

Trustee   2004-present

Profession and Affiliations: Captain, Northwest Airlines and Delta Airlines; member of nonprofit boards; first female museum Chair

Highlights during Chairmanship:

The incredibly successful *Inspiration Begins Here!* comprehensive campaign surpasses the $77 million mark, exceeding its goals

The Aviation Pavilion opens to the public in June 2016

The Museum of Flight hosts "Above and Beyond" – a groundbreaking interactive exhibit about the power of innovation, produced by The Boeing Company

The museum is headquarters for The Boeing Company's 100th-anniversary celebration in July 2016

Museum trustees embark on a new strategic vision for the decade to come

*deceased

CEO

1965
Harl V. Brackin Jr.*
Cofounder of PNAHF
See Sidebar on page 24.

1977-1991
Howard C. Lovering
Executive Director
See Sidebar on page 200.

1992-2005
Ralph A. Bufano
Executive Director/President and CEO
See Sidebar on page 250.

2005-2010
Bonnie J. Dunbar, Ph.D.
President and CEO
See Sidebar on page 270.

2010
Michael R. Hallman
Interim President and CEO
See Sidebar on page 282.

2010-
Douglas R. King
President and CEO
See Sidebar on page 326.

*deceased

Board of Trustees: 1965-2016

King County Museum of Flight Authority

King County chartered the Museum of Flight Authority board in 1985 at the request of the museum, to undertake, assist with, and otherwise facilitate or provide for the development and operation of a first-class air and space museum.

Board members since inception

Peter Anderson

Bradley Bagshaw

Richard E. Bangert
Office held: President

Robert E. Bateman
Office held: Treasurer

Arlington W. Carter Jr.
Offices held: President; Vice President

Paul R. Cressman Sr.
Office held: Secretary

Mary M. Gates
Office held: Vice President

Frank Hansen

S. C. (Sy) Iffert
Offices held: President; Vice President

B. Gerald Johnson
Office held: Of Counsel

Tom Lovejoy
Office held: Secretary

Geraldine (Geri) Lucks
Office held: Secretary

Molly T. MacFarland
Office held: Treasurer

H. Eugene McBrayer

Wells McCurdy
Office held: Treasurer

Stanley O. McNaughton
Office held: Secretary

W. Thomas Porter

William J. Rademaker Jr.

Theodore S. Roberts, MD
Offices held: Vice President; Treasurer

Bernice F. Stern

Col. Shokichi "Tok" Tokita, USAF (Ret.)
Offices held: Treasurer; Secretary

Janice Van Blaricom

Jerry A. Viscione

Judge Walter E. Webster
Offices held: President; Vice President; Secretary

Pathfinders Hall of Fame

The Museum of Flight and the Pacific Northwest Section of the American Institute of Aeronautics and Astronautics annually bestow the Pathfinder Award on those individuals with ties to the Pacific Northwest who have made significant contributions to the development of the aerospace industry. The Pathfinder Award honorees are selected by representatives of aviation and aerospace organizations and companies throughout the Northwest in the categories of Operations, Manufacturing, Flying, Engineering, Education, and At-Large.

Name	Year Inducted	Category
Edmund T. Allen	1983	Flying
William M. Allen	1984	Manufacturing
William A. Anders	2005	At-large
Michael P. Anderson	2015	At-large
William S. Ayer	2012	Operations
C. Donald Bateman	2003	Engineering
Wellwood E. Beall	1989	Engineering
Jeff Bezos	2016	At-large
William E. Boeing Jr.	2010	Education
William E. Boeing Sr.	1982	Manufacturing
Peter M. Bowers	1992	Education
Harl V. Brackin Jr.	1984	Education
Michael Carriker	2013	Flying
John Cashman	2014	Flying
Joe Clark	2008	At-large
William A. Cook Jr.	1993	Engineering
Carolyn Corvi	2011	Manufacturing
A. Scott Crossfield	1998	Eng./Flying
Joseph E. Crosson	1984	Operations
Suzanna Darcy-Hennemann	2010	Flying
Gladys Dawson-Buroker	1989	Education
Bonnie J. Dunbar	1990	At-large
Fred S. Eastman	1986	Education
Clairmont L. Egtvedt	1982	At-large
Steve Fulton	2011	Operations
Jim Galvin	2000	Operations
Ernest K. Gann	1994	At-large
Reba Gilman	2008	Education
Richard F. Gordon Jr.	2002	Flying
Louis B. Gratzer	2003	Engineering
Robert E. Hage	2007	Engineering
Elling Halvorson	2015	Operations
Thomas F. Hamilton	1982	Engineering
Elmer N. Hansen	2004	Education
David R. Hinson	2005	At-large
Thorp Hiscock	1998	Engineering

Eddie Hubbard	1988	Operations	Clayton L. Scott	1985	Flying
Joe M. Jackson	2014	Flying	Frank Shrontz	2013	Operations
Jeff Jefford	1984	Flying	Delford M. Smith	1990	Operations
Elrey B. Jeppesen	1986	At-large	Joe I. Soloy	1996	Engineering
Gene Nora Jessen	2006	Flying	John E. Steiner	1997	Engineering
Philip G. Johnson	1983	Manufacturing	Dorothy Hester Stenzel	1988	Flying
Robert R. Johnson	1986	Operations	George Stoner	1988	Engineering
A. M. "Tex" Johnston	1987	Flying	Joseph F. Sutter	1995	Engineering
James Joki, M.D.	2009	Engineering	Richard W. Taylor	1993	Flying
Bruce Kennedy	2006	Operations	Leslie R. Tower	1982	Flying
Fredrick Kirsten	1983	Engineering	Guy M. Townsend	1994	Flying
Milton G. Kuolt II	2002	Operations	A. E. "Art" Walker	1989	Flying
Clay Lacy	2010	Flying	S. L. "Lew" Wallick Jr.	1999	Flying
Robert. T. Lamson	2001	Flying	Edward C. Wells	1983	Engineering
Wesley G. Lematta	1995	Operations	Sheila E. Widnall	1996	At-large
Barbara Erickson London	1992	Flying	Noel Wien	1982	Operations
Nicholas B. Mamer	1983	Flying	Frank W. Wiley	1984	Education
Louis S. Marsh	1982	Engineering	Dr. Sam B. Williams	2004	Engineering
George C. Martin	1987	Engineering	T. A. Wilson	1991	Engineering
Bruce R. McCaw	2013	At-large	John K. Wimpress	2014	Engineering
A. Elliott Merrill	1985	Flying	Holden W. Withington	2001	Engineering
Marvin Michael	1999	At-large	Arthur C. Woodley	1987	Operations
Charles N. Monteith	1984	Engineering	Brien Wygle	2000	Flying
Barbara Morgan	2012	Education			
Dr. George E. Mueller	2012	Operations			
Alan Mulally	2015	Operations			
Robert B. Munro	1997	Operations			
Herbert A. Munter	1991	Flying			
Eric H. Nelson	2006	Engineering			
Clyde E. Pangborn	1982	Flying			
Addison Pemberton	2016	Engineering			
Maynard L. Pennell	1986	Engineering			
Raymond I. Peterson	1985	Operations			
James Raisbeck	2007	Engineering			
Robert C. Reeve	1983	Operations			
John P. Roundhill	2009	Engineering			
James S. Russell	1984	At-large			
George S. Schairer	1985	Engineering			

Photo Credits

All photos from the Museum of Flight collection (MoF) unless indicated otherwise:

Front cover: MoF, Ted Huetter

9:	MoF, Ted Huetter
12:	MoF, Bowers Collection
14b:	Howard Lovering Collection
15a:	MoF, Ted Huetter
16:	MoF, The Boeing Company Collection
17:	The Boeing Company
18a:	MoF, George Steck
18b:	The Boeing Company
19b:	Bruce McCaw Collection
21:	The Boeing Company
22b:	MoF, Gordon Williams
28:	The Boeing Company
35:	The Boeing Company
37:	MoF, Bowers Collection
38:	Jim Larsen
39:	The Boeing Company
41b:	MoF, Don Dwiggins
42-43:	Jim Larsen
46:	Jim Larsen
50:	Jim Larsen
51:	MoF, The Boeing Company Collection
52a:	Jeremy Dwyer-Lindgren
53a:	MoF, John D. Mitchell
55a:	MoF, J.N. Dilonardo
55b:	The Mira Slovak Collection of the Hydroplane Museum
58b:	MoF, Frederick A. Johnsen
61:	The Boeing Company
66a:	MoF, Steve Brackin
69:	MoF, The Boeing Company Collection
70:	MoF, Bowers Collection
71:	MoF, Bowers Collection
73:	The Boeing Company
74a:	Washington State Archives

75a-b:	The Boeing Company
77:	The Boeing Company
79:	The Boeing Company
81a:	The Boeing Company
83:	Howard Lovering Collection
86:	Vern Rutledge
87:	MoF, Ted Huetter
95:	The Boeing Company
96:	MOHAI, PEMCO Webster & Stevens Collection, 1983.10.8601.2
97:	The Boeing Company
99:	Peggy Nuetzel
102:	The Boeing Company
116:	The Boeing Company
122:	MoF, USAF
125a:	Howard Lovering Collection
132:	Howard Lovering Collection
134:	The Boeing Company
136a-b:	Peggy Nuetzel
147b:	Peggy Nuetzel
148:	Peggy Nuetzel
154:	Howard Lovering Collection
155:	Peggy Nuetzel
157a:	MoF, Bowers Collection
161:	Vern Rutledge
165:	Howard Lovering Collection
171:	Bradford Lovering
172b:	The Boeing Company
190:	Jim Larsen
191:	MoF, Heath Moffat
193a:	Peggy Nuetzel
194:	Howard Lovering Collection

206:	NASA
210a:	Martha Visk
213a:	MoF, Dennis Fleischman
215a:	MoF, Heath Moffat
216:	MoF, Ted Huetter
220:	MoF, Heath Moffat
223a:	The Boeing Company
224a:	Dennis Parks
224b:	Pacific Studio Exhibit Design and Fabrication/ Hendrickson Chase Photographics
229b:	White Rain Films
230b:	Ned Ahrens
232a:	MoF, Bill Mohn
236:	Ralph Radford
243:	Daniel Thompson
244:	Ralph Radford
245a:	MoF, Tom Beelmann
247:	Ralph Radford
252:	MoF, Layne Benofsky
255a:	MoF, Ted Huetter
255b:	The Boeing Company
256a:	University of Washington
260a:	Liz Matzelle
262a:	James Reuter
267:	MoF, Francis Zera
270:	MoF, David Waltman Studios
275:	MoF, Ted Huetter
276:	Aero-Metric Walker Division
278:	NASA
279:	MoF, Ted Huetter
280:	MoF, Ted Huetter
281b:	Dieter Zube
287a:	Adam Buchanan Photography
288:	The Boeing Company
289:	MoF, Ted Huetter
290:	MoF, Francis Zera
291:	MoF, Ted Huetter

294:	John Deane
300:	Jeremy Dwyer-Lindgren
301-303:	MoF, Ted Huetter
304:	Raisbeck Aviation High School
305-306:	Barney Britton
307:	MoF, Ted Huetter
308-310:	The Boeing Company
312:	Adam Buchanan Photography
314a-b:	Ted Huetter
315:	Adam Buchanan Photography
316:	*Seattle Met*
318:	Adam Buchanan Photography
319b:	Adam Buchanan Photography
320:	Adam Buchanan Photography
322-323a:	MoF, Francis Zera
323b:	Hayley Bostrom
324a:	MoF, Ted Huetter
325:	MoF, Ted Huetter
328:	The Boeing Company
329:	Bezos Expeditions
330a:	The Boeing Company
330b:	Blue Origin
331b-332:	MoF, Ted Huetter
334a:	MoF, Francis Zera
335b:	MoF, Ted Huetter
337:	MoF, Heath Moffat
Back flap:	Cynthia MacKenzie
Back cover:	MoF, Ted Huetter

Index